CONTINUOUS PRODUCT LANDSCAPES

CONTINUOUS PRODUCTIVE URBAN LANDSCAPES:
DESIGNING URBAN AGRICULTURE FOR SUSTAINABLE CITIES

André Viljoen
Katrin Bohn
Joe Howe

AMSTERDAM • BOSTON • HEIDELBERG • LONDON • NEW YORK • OXFORD
PARIS • SAN DIEGO • SAN FRANCISCO • SINGAPORE • SYDNEY • TOKYO

Architectural Press is an imprint of Elsevier

Architectural Press
An imprint of Elsevier
Linacre House, Jordan Hill, Oxford OX2 8DP
30 Corporate Drive, Burlington, MA 01803

First published 2005

Copyright © 2005, André Viljoen. All rights reserved.

The right of André Viljoen to be identified as the author of this work has been asserted in accordance with the Copyright, Designs and Patents Act 1988.

No part of this publication may be reproduced in any material form (including photocopying or storing in any medium by electronic means and whether or not transiently or incidentally to some other use of this publication) without the written permission of the copyright holder except in accordance with the provisions of the Copyright, Designs and Patents Act 1988 or under the terms of a licence issued by the Copyright Licensing Agency Ltd, 90 Tottenham Court Road, London, England W1T 4LP. Applications for the copyright holder's
written permission to reproduce any part of this publication should be addressed to the publisher.

Permissions may be sought directly from Elsevier's Science and Technology Rights Department in Oxford, UK; phone: +44-0-1865-843830;
fax: +44-0-1865-853333; e-mail: permissions@elsevier.co.uk.
You may also complete your request on-line via the Elsevier homepage (http://www.elsevier.com), by selecting 'Customer Support' and then 'Obtaining Permissions'.

British Library Cataloguing in Publication Data
A catalogue record for this book is available from the British Library

ISBN 0 7506 55437

For information on all Architectural Press
publications visit our website at architecturalpress.com

Typeset by Newgen Imaging Systems Pvt Ltd, Chennai, India
Printed and bound in Great Britain

Working together to grow
libraries in developing countries

www.elsevier.com | www.bookaid.org | www.sabre.org

ELSEVIER   BOOK AID International   Sabre Foundation

CONTENTS

Foreword		*ix*
Preface		*xi*
Acknowledgements		*xii*
Illustrations: credits and permissions		*xiii*
Contributors		*xv*
An introductory glossary		*xviii*
André Viljoen and Katrin Bohn		

Part One   Carrot and City: The Concept of CPULs 1

1 New space for old space: An urban vision 3
   Katrin Bohn and André Viljoen
      Continuous Productive Urban Landscapes: London in 2045 4
      Ecological Intensification 7
      London in 2045: Postscript 8

2 More space with less space: An urban design strategy 10
   Katrin Bohn and André Viljoen
      What are CPULs? 11
      Why CPULs? 12
      Where will CPULs be? 15

Part Two   Planning for CPULs: Urban Agriculture 17

3 More food with less space: Why bother? 19
   André Viljoen, Katrin Bohn and Joe Howe
      Food and urban design 21
      Why urban food? 21
      The environmental case for urban agriculture 22
      Organic urban agriculture 25
      Seasonal consumption 26
      Local growing and trading of crops 29

4 Urban agriculture and sustainable urban development 32
   Herbert Giradet
      Taking stock 33
      Energy and land-use 33
      Nutrient flows 34
      Relearning Urban Agriculture 35
      Cities as sustainable systems 38

5 Food Miles 40
   Angela Paxton
      Introduction 41
      The food miles food chain 41
      Implications of the food miles chain 44
      Forces behind the food miles 45
      Solutions 46

6 Sandwell: A rich country and food for the poor 48
   André Viljoen

7 Plan it: An inclusive approach to environmentally sustainable planning 52
   Dr Susannah Hagan

8	New cities with more life: Benefits and obstacles *Joe Howe, André Viljoen and Katrin Bohn*	56
	Socio-cultural benefits	57
	Economic benefits	57
	Health benefits	59
	Obstacles to urban agriculture	60
9	The economics of urban and peri-urban agriculture *James Petts*	65
	Introduction	66
	Micro-economic aspects	68
	Motives for UPA	68
	Supply	69
	Market entry, demand, and prices	71
	Macro-economic aspects	72
	Policy conclusions	73
10	Changing consumer behaviour: The role of farmers' markets *Nina Planck*	78
11	The social role of community farms and gardens in the city *Jeremy Iles*	82
	What is a community garden or a city farm?	83
	The benefits of a community garden or city farm project	83
	Policy on food production and sustainability	85
	Encouraging an existing or new project	86
	A vision for the future	87
	The Federation of City Farms and Community Gardens	87
	Notes	88
12	Recycling systems at the urban scale *Dr Margi Lennartsson*	89

Part Three   Planning for CPULs: Open Urban Space — 93

13	Food in time: The history of English open urban space as a European example *Joe Howe, Katrin Bohn and André Viljoen*	95
	Where we live is where we grow	97
	Industrial revolution and suburban utopias: the divorce of cities and food production	98
	Urban food and conflict	101
	Urban rebuilding and urban food decline	104
	Revival and diversification of urban food growing	105
	Urban agriculture and sustainability	106
14	Food in space: CPULs amongst contemporary open urban space *Katrin Bohn and André Viljoen*	108
	CPULs for European cities	109
	The open urban space atlas	109
15	Designs on the plot: The future for allotments in urban landscapes *Professor David Crouch and Richard Wiltshire*	124
	Allotments as (open) green space	125
	Allotments as urban landscape	127
	Allotments as negotiated communities	129
	The plot for designers	130

Part Four   Planning for CPULs: International Experience — 133

16	Urban agriculture in Havana: Opportunities for the future *Jorge Peña Díaz and Professor Phil Harris*	135
	The challenge	136
	People's and government's response	137

	Rural agriculture and urban horticulture in Cuba	138
	Urban and peri-urban agriculture in Havana	140
	Havana: antecedents and current development	140
	The experience of the 1990s	142
	Conclusions	143
	References	145
17	Cuba: Laboratory for urban agriculture	146
	André Viljoen and Joe Howe	
	The spatial characteristics of Cuban urban agriculture	147
	Urban agriculture at the city scale	149
	The urban agriculture site	153
	Infrastructure and urban agriculture	188
	Urban agriculture in residential areas	189
	Organic urban agriculture	190
18	Urban and peri-urban agriculture in East and Southern Africa: Economic, planning and social dimensions	192
	Dr Beacon Mbiba	
	Research Patterns since the 1970s: regional and subject focus	193
	Summary of the context of urban and peri-urban agriculture in africa in critical research areas	195
	Actors and institutions	197
	Conclusion	198
19	Moulsecoomb: Discovering a Micro-PUL	200
	André Viljoen and Katrin Bohn	
20	Allotments, plots and crops in Britain	206
	H.F. Cook, H.C. Lee and A. Perez-Vazquez	
	The organisation of allotments	207
	Practical production issues	210
	The economic implications of UPA	214
	Conclusions	215
21	Urban food growing: New landscapes, new thinking	217
	Simon Michaels	
	Food growing in urban areas	218
	The landscape character of urban food growing projects	218
	Conclusion	220
22	Permaculture and productive urban landscapes	221
	Graeme Sherriff	
	Permaculture	222
23	Utilitarian dreams: Examples from other countries	229
	André Viljoen	
	Delft, The Netherlands	230
	Kathmandu Valley, Nepal	230
	Gaborone, Botswana	231
	Another model	233

Part Five   Carrot and City: Practical Visioning **237**

24	New space for old cities: Vision for landscape	239
	Katrin Bohn and André Viljoen	
	Size	240
	Sense of openness	242
	Local interactions	244
	Urban nature	245
	Persistent visual stimulation	249

25	More city with less space: Vision for lifestyle *Katrin Bohn and André Viljoen*	251
	Variety of occupation and occupants	252
	Economic return from land-use	253
	Inner-city movement	260
	Environmental delight	262
26	More or less: Food for thought *André Viljoen and Katrin Bohn*	265
Contacts		*270*
Index		*272*

FOREWORD

With a vision and a strategy the 21st century city will be green, a healthy place for all and will generate zero net pollution. This book offers a vision and a strategy.

Productive urban landscapes have two huge challenges to address: CO_2 emissions are projected to increase by two-thirds in the next 20 years, and as the global food production increases so does the number of people going hungry, with the number of urban hungry soaring.

The symbiotic relationship between a productive landscape and the human settlement system is as old as civilization. During the past 200 years that millennium-old positive relationship deteriorated into a further and further separation of town and landscape. The good news is that during the past quarter-century the agriculture industry has turned a corner towards greater integration with our modern cities.

One of the earliest archeological evidences of CPULs (4,000 years ago) are the semi-desert towns of Persia. Underground aqueducts brought mountain water to oases where intensive food production was conducted, substantially based on the use of urban waste within the settlement.

A marvellous example in history is Machu Picchu in Peru. The Spaniards did not discover this nutritionally self-reliant city for 100 years. Scarce water was reused time and again, step-by-step down the mountain. Biointensive vegetable beds were designed to catch the afternoon sun and stretch the season. Water and land crops were brought together to resist the frequent mountain frost. There are many such stories from all corners of the earth.

The industrial revolution brought the railroads, chemical fertilizers, petroleum fuel, tinned food and refrigeration and a separation of the food system from where we live. Socially this converted to the creation of the 'city slicker' and of the 'country bumpkin'. Ecologically it brought many dreadful patterns of sickness, worst today in the Himalayan city of Katmandu.

Our current industrial and agricultural systems transport by ship, rail, truck and plane over 80 percent of all extracted natural resources to four percent of the Earth's land and on that urban four percent convert over 80 percent of it to waste and pollution. The interpretation *'waste is food'* enables us to conceive of operating systems that utilize waste (heat, sewage, waste-water runoff, organic solids, construction debris, etc.) to green the city and feed the urban population of the globe by closing now-open nutrient cycles.

In the later 1970s there emerged reports of a resurgence of agriculture in the city from (alphabetically) Bogota, Dubai, Lusaka, Madrid, Manila, Moscow, New York, Vancouver and from many corners of the globe. A United Nations survey of 20 countries around-the-world and library research conducted in 1991-1993 concluded that there was the beginning of a new urban-based food system evolving worldwide.

This book is a 21st century breakthrough in defining an urban design/planning conceptual approach to re-incorporating a productive landscape, including agriculture, into the human settlement (CPULs). As reported in the chapter 'Food in Time' in the previous hundred years there were several such models created including famously: Le Courbusier, Paul & Percival Goodman, Ian McHarg, Louis Mumford, and Frank Lloyd Wright. We have both history and great creative minds to guide our hands to this gigantic task.

Agriculture, reaching from fish farming to ornamental shrubs, is moving to the mostly urban market and becoming less centralized in a few corporations. The potential for CPULs it seems to me is eminently workable based on two characteristics of 21st century cities: constant renewal and constant de-densification.

Cities today are constantly renewing themselves. Yesterday's factory sites, shopping malls, and housing estates are collapsing and standing idle for a decade or two or three. These sites, which are idle on an interim basis, are a foundational element in the locally-based food system and the ecologically sustainable (green) city.

FOREWORD

The emerging 21st century city can be identified as 'the Edgeless City'. The concepts of city boundary, greenbelt, and suburb are all obsolete. The city that was focussed on the river, the seaport, the railyard, and the limited access highway intersection are all obsolete. Cities are becoming formless, edgeless and seemingly endless. In Africa the city extends from Abidjan to Lagos, in Asia from Kobe-Osaka to Tokyo-Chiba, in North America from Portland Maine to Norfolk Virginia, and in Europe from Barcelona to Genoa.

Once enlightened by the CPUL concept our eyes can see possibilities everywhere: the waste heat from supermarket refrigeration is a source of energy for food production, flood plains are productive if producing crops and costly if used for housing, fruit and vegetable production on rooftops saves heating and cooling costs, reduces air pollution and enables fresh cuisine, a security fence is a potential for productive and ornamental vines.

Greening the 21st century city will improve our health, stabilize our economy and bring us all closer together as we meet in the garden.

Jac Smit, AICP

PREFACE

This book is intended to contribute to the ongoing debate about the future shape of cities.

Supported by emerging international research, it presents a vision for integrating Continuous Productive Urban Landscapes (CPULs) into existing and future cities. CPULs are urban spaces combining agricultural and other landscape elements within a strategy of continuous, open space linkages.

The book focuses on design and planning questions raised by CPULs and examines the various qualities CPULs can bring to the urban fabric. Chapters by Katrin Bohn, André Viljoen and Joe Howe present the case for CPULs, exploring the situation today, the historical context and proposals for CPUL design strategies. A series of underpinning chapters, written by specialists, develop and expand upon these issues.

Urban agriculture within CPULs, integrated into individual cities, can contribute to more sustainable food production and open space management. If the design potential of CPULs is to be realised, it is necessary to understand the arguments supporting urban agriculture.

CPULs will form part of an urban infrastructure and as such, their adoption implies embarking on a long-term development strategy which is equally applicable to established and emerging cities. The book explores different ways of implementing CPULs, using both visionary proposals and practical experience to support the argument for their adoption.

André Viljoen, Katrin Bohn and Joe Howe

ACKNOWLEDGEMENTS

Three of us have worked on the main text for this book, and a number of individuals and institutions have supported us along the way. Particular thanks go to those who have contributed specialist chapters, always efficiently and on time. Their input has been critical in underpinning the concept of CPULs.

Katrin Bohn and André Viljoen would like to thank Guisi Alcamo and Martina Oppel, who assisted with initial research. Eva Benito, Katja Schäfer, Lucy Taussig and Kabage Karanja who have helped to develop drawings and illustrations. We especially appreciate the insightful comments Kim Sorvig and Jac Smit made about earlier drafts. The Faculty of Arts and Architecture at the University of Brighton has supported this project from its inception with the Centre for Research and Development and the School of Architecture and Design funding research. In particular, we would like to thank Anne Boddington, Prof. Bruce Brown, Prof. Jonathan Woodham and Sean Tonkin. At the Architectural Press, Alison Yates, Liz Whiting and Catharine Steers for their help and patience. And μ, Alma and Bertolt for, amongst many things, visiting the Peckham Farmers Market every Sunday.

Moreover, André would like to thank Jorge Pena Diaz, from the School of Architecture at City University José Antonio Echeverría (CUJAE) in Havana for facilitating the initial field trip to Cuba, and our ongoing work in Cuba and the UK. In Cienfuegos, Prof Padron Padron and Prof. Socorro Castro who made our visit so fruitful. Tom Phillips for joining in with the urban agriculture adventure and our ongoing research at the Peckham Experiment. Eddie Edmundson and Yania Lucas from the British Council in Havana for their ongoing assistance. In particular I would like to acknowledge support for my second field trip to Cuba with Tom Phillips, and for arranging Yuneikys Villalonga's assistance in Havana. And a special word of thanks to all the urban farmers, administrators and planners in Cuba who answered our questions and let us photograph, draw and learn about their organoponicos, which have now been running successfully for over 10 years. Rene Van Veenhuizen, from RUAF (Resource for Urban Agriculture and Forestry) for assistance and support. Angela Blair from the Rowley, Regis and Tipton primary care trust for introducing us to the Sandwell food mapping project. Warren Carter from the Moulsecoomb Forest Garden and Wildlife Project in Brighton, for providing access to the project. The Royal Institute of British Architects, Modern Architecture and Town Planning Trust, for supporting the project initially with a research award. Robert Mull and Prof. Mike Wilson at London Metropolitan University's School of and Architecture Spatial Design, and the Low Energy Architecture Research Unit, where the project originated.

Katrin would like to thank Hans Gebauer, Janet Rudge, Abby Taubin, Jens Weber and Harry and Inge Bohn for reading, discussing, photographing and encouraging our work. And the Bauhaus-Universität Weimar, Germany, where the project somehow originated as well.

Joe Howe wishes to acknowledge the support of the Economic and Social Research Council (ESRC) for funding research undertaken during 2000–2001, which was concerned with urban agriculture and land-use regulation in metropolitan areas of the UK. This research has fed into the book.

ILLUSTRATIONS: CREDITS AND PERMISSIONS

All images, unless otherwise noted, are copyright Bohn & Viljoen Architects.

Fig. 1.1 Image created by Andre Viljoen, with extracts from the Geographers' A-Z M25 main road London Map as an underlay with permission from Geographers A-Z Map Company Ltd.

Fig. 3.1 redrawn by Bohn & Viljoen Architects, from Fuel's Paradise: Energy options for Britain by Peter Chapman (Penguin Books 1975) © Copyright Peter Chapman, permission from Penguin Books.

Fig. 3.2 assembled by Bohn & Viljoen Architects from Building a sustainable future: Homes for an autonomous Community, General Information Report 53, B. Vale and R. Vale © Crown Copyright Reproduced with permission of the Controller of HMSO and the Queen's Printer for Scotland. And from Energy Policy, Vol 27, Klaas Jan Kramer, Henri C. Moll, Sanderine Nonhebel, Harry C. Wilting, 'Greenhouse gas emissions related to Dutch Food consumption', Pages 203–216, 1999 with permission from Elsevier Science.

Fig. 3.3 assembled by Bohn & Viljoen Architects from, Bohn & Viljoen research. And from Energy Policy, Vol 27, Klaas Jan Kramer, Henri C. Moll, Sanderine Nonhebel, Harry C. Wilting, 'Greenhouse gas emissions related to Dutch Food consumption', Pages 203–216, 1999 with permission from Elsevier Science. And from DETR HMSO (2000) English House Condition Survey 1996. Table 8.7 pg 103, Production of CO2 by tenure, figures extracted for use in chart. © Crown Copyright Reproduced with permission of the Controller of HMSO and the Queen's Printer for Scotland.

Fig. 3.4 This was originally published in 'The fires of culture' Steinhart, C. and Steinhart, J. 1974, Belmont California: Wadsworth Publishing Company.

Fig. 3.5 created by Bohn & Viljoen Architects based on material in Leach, G. (1976) *Energy and food production*, Institute for Environment and Development., IPC Science and Technology Press.

Fig. 3.6 created by Bohn & Viljoen Architects based on material in Leach, G. (1976) *Energy and food production*, Institute for Environment and Development., IPC Science and Technology Press.

Fig. 3.7 produced by Bohn & Viljoen Architects based on material in Lampkin, N.H. and Padel, S. (1994) *The Economics of Organic Farming*, CABI publishing, based on the original source, Murphy, M.C. (1992) *organic farming as a Business in Great Britain*. Agricultural Economics Unit, University of Cambridge, Cambridge.

Fig. 3.8 based on material in Kol, R., Bieiot, W. and Wilting, H.C. (1993) *Energie-intensiteiten van voedingsmiddelen*, Energy and Environmental Sciences Department (IVEM) State University of Groningen, Netherlands.

Fig. 5.1 and 5.2 © Crown Copyright Reproduced with the permission of the Controller of HMSO and the Queen's Printer for Scotland.

Fig. 6.1 published with permission from Rowley Regis and Tipton primary care trust and maps. Reproduced by permission of Ordnance Survey on behalf of the Controller of Her Majesty's Stationary Office, © Crown Copyright 100040510.

Figs 9.1 and 9.2 by permission of Sustain.

Fig. 9.3 by permission of S. Gertstle.

Figs 13.1–13.7 are all Crown Copyright, Reproduced by permission of the Controller of Her Majesty's Stationary Office, and produced with the permission of the trustees of the Imperial War Museum, London.

Fig. 17.2 drawn by Bohn & Viljoen Architects, based on site surveys and a report prepared by the University of Cienfuegos.

ILLUSTRATIONS

Fig. 17.3   drawn by Bohn & Viljoen Architects, based on site surveys and a report prepared by the University of Cienfuegos.

Figs 18.1–18.3   are all by permission of Dr. B. Mbiba.

Fig 19.5   by permission of Antonia Faust.

Figs 20.1–20.3, 20.5 and 20.6 by permission of A Perez-Vazques.

Figs 22.1–22.3   by permission of G. Burnett.

Fig. 22.4   by permission of G. Sherriff.

Fig. 22.5   by permission of G. Burnett.

Figs 22.6–22.8   by permission of G. Sherriff.

Fig. 23.1   by Jan Hiensch for the Urban Agriculture Magazine. Urban Agriculture Magazine Number 4, July 2001. Based on the original illustrations by D. Boyd, IIUE.

Fig. 23.2   by Jan Hiensch for the Urban Agriculture Magazine. Urban Agriculture Magazine Number 4, July 2001. Based on the original illustrations by I. Boyd and K. Weise, PAHAR.

Fig. 23.3   by Jan Hiensch for the Urban Agriculture Magazine. Urban Agriculture Magazine Number 4, July 2001. Based on the original illustrations by Prof. A C Mosha and B. Cavric.

Fig. 23.5   by Jan Hiensch for the Urban Agriculture Magazine. Urban Agriculture Magazine Number 4, July 2001. Based on the original illustrations by M.D. Kitilla, and A. Mlambo.

Fig. 23.6   by Jan Hiensch for the Urban Agriculture Magazine. Urban Agriculture Magazine Number 4, July 2001. Based on the original illustrations by P. Mishev and A. Yoveva, SWF, Sofia, Bulgaria, Architect.

Fig. 24.1   created by Bohn & Viljoen Architects, using a photograph produced by Simmons Aero Films Limited, as an underlay.

Plate 1   by permission of Tom Phillips.

Plate 6   created by Bohn & Viljoen Architects, using an extract from the Phillips Maps, London and M25 navigator map as a underlay.

Plate 7   LeisurESCAPE London Southwark detail map, original image created by Bohn & Viljoen Architects, using an extract from the an ordnance survey map as underlay, reproduced by permission of Ordnance Survey on behalf of the Controller of Her Majesty's Stationary Office, © Crown Copyright 100040510.

Plate 8   LeisurESCAPE London Southwark detail aerial view collage, original image created by Bohn & Viljoen Architects, using an extract from 'London: the photographic atlas' permission requested from Harper Collins Illustrated.

CONTRIBUTORS

Katrin Bohn is an architect and senior lecturer at the School of Architecture and Design at the University of Brighton, where she runs a design studio with Andre Viljoen. Within her urban design research, she has developed several architectural and landscape proposals, mostly centred around CPULs. Recent live projects relating to landscape and ecological building include the CUE Eco House in London (with the Low Energy Architecture Research Unit at London Metropolitan University) and proposals for community landscapes in Southwark, London.

Dr Hadrian F. Cook is a member of the Agroecology Research Group, at Wye Campus, Imperial College. Main Environmental Research Interests are: soil amendment using organic wastes; hydrology of grazing marshes and watermeadows; protection of surface and groundwaters from agrochemical pollution; protection policy development for soil and water and environmental history.

David Crouch is Professor of Cultural Geography, Tourism and Leisure at the University of Derby, Visiting Professor Geography and Tourism, University of Karlstad Sweden; author of several publications related to Allotments including (with Colin Ward) *The Allotment: its landscape and culture* (Faber and Faber/Five leaves Press 1988, 1994, 1997, 2001). He has contributed to a number of reports for NGO's and government, as well as producing for BBC2 TV, 'The Plot' in 1994.

Herbert Giradet is a social anthropologist and cultural ecologist now working as a writer, consultant and film maker. His main focus in recent years has been the sustainable development of cities and contemporary lifestyles. He is widely published, a prolific documentary maker and has been invited to work on sustainability in many countries around the world. Herbert is the recipient of the UN Global 500 Award for Outstanding Environmental Achievement, an honorary Fellow of the Royal Institute of British Architects and Chair of the UK Sustainability Alliance and Schumacher Society.

Dr Susannah Hagan trained as an architect at Columbia University and the Architectural Association. She is Reader in Architecture and head of the MA Architecture: Sustainability, at the University of East London and also teaches on the AA's Environment and Energy Graduate Programme. Her book, *Taking Shape*, explores the relationships between the built and natural environments.

Phil Harris is Professor of Plant Science at Coventry University and International Research Consultant for the Henry Doubleday Research Association. Current research interests include tropical crop development, 'organic' or sustainable agriculture, forestry, and relevant techniques of plant biotechnology, often related to overseas development. Research and consultancy activities in sustainable agriculture and forestry have involved work in Bangladesh, Brazil, Cape Verde, China, Cuba, Ghana, India, Japan, Jordan, Kenya, Oman, Sierra Leone, South Africa and Spain.

Dr Joe M. Howe is a senior lecturer at the University of Manchester's School of Planning and Landscape. His research focuses on the relationship between sustainability and planning. This has encompassed work on urban food growing and recently on the relationship between water management and land use planning and management. He has advised numerous Government bodies (including DEFRA, ODPM, DTI and the Treasury) and NGOs on water and land use management issues.

Jeremy Iles's career in the environmental sector has included roles at Friends of the Earth (Transport Campaigner), the London Wildlife Trust (Director), overseas work in Bangladesh and Eritrea as Field Director for VSO, and as a Regional Manager on the National Cycle Network Project at Sustrans. He took up the role as Director of the Federation of City Farms and Community Gardens in autumn 2000.

Dr Howard Lee's interest is in sustainable agriculture. His contribution to this book stems from his work at the Agroecology Research Group, Wye Campus, Imperial College. His main research areas are: managing the water resource base in agroecological water catchment zones; nitrogen dynamics in farming systems and environmental impact; the environmental impact of organic waste management on farms and in the community and the use of geographical information systems to predict the environmental impact of farming.

Dr Margi Lennartsson is Head of the International Research Department at HDRA responsible for scientific research activities of the Association. HDRA is a registered charity involved in research, advisory work and

CONTRIBUTORS

promotion of organic food, farming and gardening. The aim of HDRA's research programme is to develop the techniques used in organic agriculture and to advance the knowledge of organic production systems, focusing on commercial organic horticulture and domestic gardening in temperate areas and on small scale, resource poor systems in developing countries.

Dr Beacon Mbiba is the Co-ordinator of the Urban and Peri-Urban Research Network Peri-NET, South Bank University, London. Research interests include local level development planning, land transformations and sustainable human settlements. He has published a number of articles on urban and peri-urban agriculture and has taught at the University of Zimbabwe and the University of Sheffield.

Simon Michaels is a landscape architect, urban designer and environmental planner. He works as an independent consultant, and is a director of f3, the UK's Foundation for Local Food Initiatives. He also runs Environment Go, an internet information service for environment professionals, and advises on internet strategies for environment sector organisations.

Angela Paxton works at the Centre for Alternative Technology in Wales and is currently setting up a community scheme with community composting, organic nursery and demonstration gardens. She is author of *The Food Miles Report, The Food Miles Action Pack* and *A Feast Too Far*, all available from Sustain.

Jorge Peña Diaz is an architect, lecturer and researcher at the Centre for Urban Studies, School of Architecture, City University José Antonio Echeverría (CUJAE) in Havana. His research has focused on the integration of urban agriculture in Havana. He has been a visiting academic at the University of Brighton and has collaborated with a number of international research and academic partners.

James Petts has a background in economics and the food industry. He is currently working for the Countryside Agency in England and has previously worked for Sustain (where his chapter was written) as a policy officer, co-ordinating the East London Food Futures project which aimed to initiate local food projects and develop a network of projects in East London, as well as working on a number of other projects to develop a more sustainable and just food economy.

Nina Planck opened London's first farmers-market in 1999. Today London Farmers Markets operates a number of weekly farmers-markets, which provide home-grown food to city dwellers and crucial income for family farms in the Southeast. Her goals for the farmers-markets are more food produced within the M25 (London's ringroad), and more organic food. She is the author of the Farmers-Market Cookbook and was an advisor to the Prince of Wale's Rural Task Force.

Graeme Sherriff graduated from Keele University with an MA in Environmental Law and Policy. His MA dissertation, based on an extensive survey of food growing projects in the UK, looked at permaculture and its relevance to sustainable agriculture. After graduating, Graeme worked on practical environmental improvement and community building projects as part of Groundwork and then became involved with research on related subjects at the University of Manchester.

Jac Smit, between the ages of 12 and 22, held jobs in a diverse range of peri-urban agriculture fields including production, processing and sales of: poultry, vegetable, livestock [goat, cow and horse], orcharding [apple, cherry and maple], and ornamental horticulture. He acquired a first degree in agriculture and a master's degree from Harvard University in city and regional planning. As project manager, technical director and principal planner he incorporated agriculture into the regional plans for Baghdad, Calcutta, Chicago, Karachi, and the Suez Canal Zone. In the early 1990s he carried out a worldwide study for UNDP to define the current and potential role of urban agriculture, which was launched at the Global City Summit in 1996. Since 1992 he has been the president of the Urban Agriculture Network and is a founding member of the global Resources Center for Urban Agriculture that has eight information centers on the five continents. He is a frequent conference presenter and is frequently published in a wide diversity of media.

André Viljoen is an architect and senior lecturer at the School of Architecture and Design at the University of Brighton, where he is undergraduate architecture course leader and runs a design studio with Katrin Bohn. Previously he was Deputy Director of the Low Energy Architecture Research Unit, based in the School of Architecture and Spatial Design at London Metropolitan University. He has participated in a number of European research studies for low energy buildings and his work in urban agriculture and urban design stems from an interest in architecture and environmental issues. Recent research and practice has concentrated on the design implications of the integration of urban agriculture into urban landscape strategies.

Arturo Perez Vazquez is completing a PhD in the Department of Agricultural Sciences Wye Campus, Imperial College. Its subject is the future role of allotments in England as a component of urban agriculture. He has received an Agropolis Award from the Cities Feeding People Program run by Canada's International Development Research Centre (IDRC).

Richard Wiltshire is a senior lecturer in Geography at King's College London and Research Officer for QED Allotments Group, a Local Agenda 21 Initiative in Dartford. His recent research focuses on the development of allotments and community gardens in Japan, and he is the co-author (with David Crouch and Joe Sempik) of *Growing in the Community* (Local Government Association, 2001) and *Sustaining the Plot* (Town and Country Planning Association, 2001) with David Crouch. He is a Steering Group member for the Allotments Regeneration Initiative.

AN INTRODUCTORY GLOSSARY

LANDSCAPE AND ENVIRONMENTAL CONCEPTS

Continuous Productive Urban Landscapes (CPULs, pronounced See Pulls) are

- the theme of this book, and do not yet exist in cities.
- a coherently planned and designed combination of Continuous Landscape and Productive Urban Landscape.
- open urban landscape.
- productive in economical and socio-cultural and environmental terms.
- placed within an urban-scale landscape strategy.
- constructed to incorporate living and natural elements.
- designed to encourage and allow urban dwellers to observe activities and processes traditionally associated with the countryside, thereby re-establishing a relationship between life and the processes required to support it.

Continuous landscape is

- a current idea in urban and architectural theory, short sections of which have been established in various cities.
- a network of planted open spaces in a city which are literally spatially continuous, such as linear parks or inter-connected open patches, sometimes referred to as an ecostructure or green infrastructure.
- virtually car-free, allowing for non-vehicular movement and encounters in open urban space.
- an alternative use of open urban space if compared to existing spatial qualities of roads and dispersed patches of used and unused open urban space.
- an enormous walking landscape running through the whole city.

Productive urban landscape is

- open urban space planted and managed in such a way as to be environmentally and economically productive, for example, providing food from urban agriculture, pollution absorption, the cooling effect of trees or increased biodiversity from wildlife corridors.

Urban agriculture is

- agriculture which occurs within the city.
- in most cases high yield market gardens for fruit and vegetable growing.
- found on the ground, on roofs, facades fences and boundaries.
- if economic conditions are difficult, likely to include small animals.
- developing to include aquaculture (fish production).

Peri-urban agriculture is

- agriculture occurring on the urban-rural fringe, or within peripheral low-density suburban areas.
- similar to urban agriculture, although the size of sites is often larger.
- UPA refers to a mix of urban and peri-urban agriculture.

Ecological footprint is

- the theoretical land and sea area required to supply the resources needed to sustain an entity (city, person, organism, building, etc.)
- partially reinstated in urban areas if CPULs are successfully implemented

Ecological intensification is

- an increase in local urban biodiversity.
- a compensation for an existing loss of biodiversity found in many urban areas.
- one of the benefits of CPULs.

Vertical and Horizontal intensification is

- increasing the number of activities or uses of a particular piece of land by overlaying one above the other.
- for Vertical intensification: usually achieved by constructing a building or series of platforms on the site, some or all of which may be used for vegetation or agriculture.
- for Horizontal intensification: applied directly on the ground by increasing the number of uses for a particular piece of land at different times and by providing access and spaces for a variety of activities and uses.
- also found in market and home gardens, layers consisting of tall to small trees, shrubs and bushes, field crops, root crops, water crops, plus fish, poultry and rabbits.
- possible, by planting on fences and walls of all types.
- multicropping, season extension, rooftop use, basement mushroom growing and floating islands (Kashmir and Burma).
- an important feature of CPULs.

TYPES OF URBAN AGRICULTURE

Sprawl is

- the expansion of cities outwards, generally at suburban densities and reliant on the car for access to work, culture and recreation.

Brownfield sites are

- plots of land which were previously occupied by industry, e.g. factory sites.
- often contaminated by chemical waste products from their previous industrial use.
- generally considered to be a primary source of new land for development in existing, and especially post industrial cities.
- currently being used as sites for new urban buildings.
- suitable for CPULs, if appropriate soil conditions exist, or if contaminated soil is treated or renewed in areas where edible crops will be grown.

The CPUL model challenges the notion that all brownfield sites should be built upon, but does not challenge the principle that all land should be used to maximise its sustainable return.

Greenfield sites are

- pieces of land which have never been built on before, e.g. farmland, forests, parks and wilderness.
- often the preferred sites for new suburban development (sprawl).

Allotments are

- found in the United Kingdom.
- for the non-commercial growing of food and flowers, rented to individuals by local authorities.
- typically 250 m^2 in area.
- clustered together in groups, a small allotment site having about 20 plots and a very large site containing several hundred plots.
- avilable from local authorities to individuals who request them.

Schrebergärten are

- found in Germany.
- similar to allotments, but not only for food growing.
- also used as weekend leisure gardens, often with a small summer house.
- with different names, spread all over Europe, further east used more for food growing.
- generally bigger than allotments but with similar situation and organisation.

Parcelas and Huerto intensivos are

- found in Cuba
- similar to allotments, though an individual plot may be larger and may be farmed by a family or group of individuals.

Organiponicos (popular and de alto rendimiento) are

- high-yield urban commercial market gardens, found in Cuba.
- based on the Chinese bio-intensive model.
- producing food for sale to the public, using raised beds and intensive organic farming methods.

Autoconsumos are

- similar to Organiponicos, but located within state enterprises with the main purpose of supplying food for employees; their yield is less than for an Organoponico.

Community gardens are

- managed and used by local communities or neighbourhoods for recreation and education.
- sometimes found on unused or abandoned urban sites, or in grounds of public buildings, e.g. public housing, hospitals, retirement homes.
- often have a small building for use by the community, in particular children and disadvantaged groups.

City farms and urban farms are

- similar to a community garden, but with animals, usually horses, goats, sheep, pigs, ducks and chickens. Their significance is educational rather than productive, although a limited quantity of produce may be generated.

Home gardens/back gardens are

- plots found behind detached or semi-detached houses, traditionally used for leisure and/or vegetable growing.

FOOD

Food security

- is defined as giving populations both economic and physical access to a supply of food, sufficient in both quality and quantity, at all times, regardless of climate and harvest, social level and income (WHO Europe, 2000).

Seasonal and local food

- is basic or core, backed up or supplemented by the globally based food system.
- is dependant on local climate and conditions for growing period, and uses the minimum of artificial stimulants, i.e. a greenhouse might be used to extend the growing season, but heating and manufactured growth promoters are avoided.
- can contribute to a reduction in imported food.
- is not going to replace all imports of fruit and vegetables.
- is an alternative to a multitude of semi-ripe imported crops currently available in developed countries.

Organic food

- is grown without the use of artificial fertilisers and pesticides.
- can contribute to reducing urban waste creating a circular urban metabolism by using compost produced from organic, domestic and farmyard waste.
- is a feature of CPULs.

Supermarket food

- relies on the importation of crops from around the world to provide the maximum choice for consumers.
- cities provide different environmental, social and economic contexts for CPULs.

Box schemes are

- a commercial service delivering a selection of organic fruit, vegetables and sometimes other products to individual homes or to a neighbourhood depot for collection.

Food miles are

- the distance food has been transported between primary production and consumption.

ECONOMIC TERMS

Factors of production

- the entities required to produce a good or service, often thought of as land, labour, and capital, but more recently widened to include human capital, social capital, physical capital, environmental capital, and financial capital.

Household

- a group of people who live in the same dwelling and share common housekeeping and eating arrangements.

Opportunity cost

- the nearest alternative cost of a factor or activity.

Fungible income

- the indirect income gained from the substitution of market-bought produce.

Formal/informal

- distinction between recorded commercial activities (formal) and unrecorded semi- or non-commercial activities (informal).

Shoe leather costs

- the incidental costs associated with travelling to and from locations of work or activity.

Barriers to entry

- the obstacles preventing new businesses entering a market.

Usufruct

- the use of land not owned by the users themselves.

Utility

- the usefulness of a product or service, the satisfaction which a consumer gets from a good or service he or she has bought, or the way in which a good or service contributes to a consumer's welfare (Collin, 2003).

Elasticity

- the responsiveness of either demand or supply to changes in price or quantity.

Externalities

- the external economic, social and environmental costs and benefits of an activity.

Food access

- both geographical and monetary degree of access to food, determined by income, supply, transport, public provision, storage, and other factors.

Acronyms

CAP – The Common Agricultural Policy (EU)

FAO – Food and Agriculture Organisation of the United Nations

GDP – Gross Domestic Product

GNP – Gross National Product

PPG – Planning Policy Guidance (UK)

UA – Urban Agriculture

UNDP – United Nation's Development Programme

UPA – Urban and Peri-urban Agriculture

WHO – World Health Organisation of the United Nations

REFERENCES

Collin, P. H. (2003). *Dictionary of Economics*. Bloomsbury Publishing PLC, London.

WHO Europe (2000). *WHO food and nutrition action plan*. WHO Europe.

PART

ONE

CARROT AND CITY: THE CONCEPT OF CPULs

1

NEW SPACE FOR OLD SPACE: AN URBAN VISION

Katrin Bohn and André Viljoen

Continuous Productive Urban Landscapes (CPULs) started to develop in London around 2005. Forty years later they were everywhere, having reached a maturity that enables us to study their success in relation to the very initial design intention.

When first mentioned around the year 2000, there was no precedent for CPULs anywhere in the world, though various attempts had been successfully started to integrate both continuous landscapes and urban agriculture into cities. As a strategy based on the genius loci of place, it was clear early on, that CPULs would have to be developed individually for each country and for each city, and that any manifesto would only provide a general framework and vision . . .

CONTINUOUS PRODUCTIVE URBAN LANDSCAPES: LONDON IN 2045

On a sunny summer Sunday morning, a CPUL in London resembles Brighton beach or Hyde Park. People have left their homes to enjoy fresh air, spaciousness and various activities of the adjacent CPUL. In the park-like areas, they do morning exercises, sit on large blankets having breakfast, sunbathe, repair their bikes or read their papers and palm top news. Children are running and playing with their friends on the grassy land between the agricultural fields or in the small canals built to water or drain the fields. Although there may be more apartment dwellers around, the CPUL is fairly evenly used by people from every housing type. One has to note, that the popularity of high-rise and dense buildings around CPULs has increased enormously since the building facades became gardens, the garden's landscapes and the views from the towers offer a visual feast.

The three local farmers' markets situated at the CPUL edge start selling fresh food, being busier today than during weekdays, when only two of them are open at any one time. (London-wide, there are now about 150 farmers' markets.) Ice cream and fresh fruit juice vendors are setting up their stalls around the main CPUL routes. The cafés and restaurants bordering the CPUL put their chairs and tables out, the smell of coffee and fresh bread blows over the fields. Tennis players exchange first balls on the nearby tennis court. Close to it, the bowls and boules groups are getting together. The various canopied outside offices, situated in the quieter areas of the CPUL, are less busy today. With their fixed-seating laptop plug-ins or their workbenches, they are now used by kids playing computer games or making aeroplane models. (During weekdays, children use computer playgrounds or youth workshops, often located closely to CPULs so as to allow safe access and the use of outside space.)

Most of the commercial farmers celebrate the weekend and the low gates to their fields are now shut, but instead other fields are busy with allotment growers and communal farm projects. There are numerous allotments within this CPUL, but they do not threaten to take over the larger, more generous urban agricultural fields: the number of allotments in London quickly stabilised, once everybody wanting to grow their own food was supplied with one. Often, the produce from the land is sold straight off the fields via one of the various small kiosks that are situated within the CPUL allowing farmers to weigh, price and record produce appropriately. For the past 20 years, since about 2025, air pollution has no longer been an issue and ground contamination is being cleared through systematic soil treatment and continuous planting. The organic produce on offer is therefore in high demand, making the markets and kiosks a bustling counterpoint to the tranquillity of this Sunday morning.

NEW SPACE FOR OLD SPACE: AN URBAN VISION

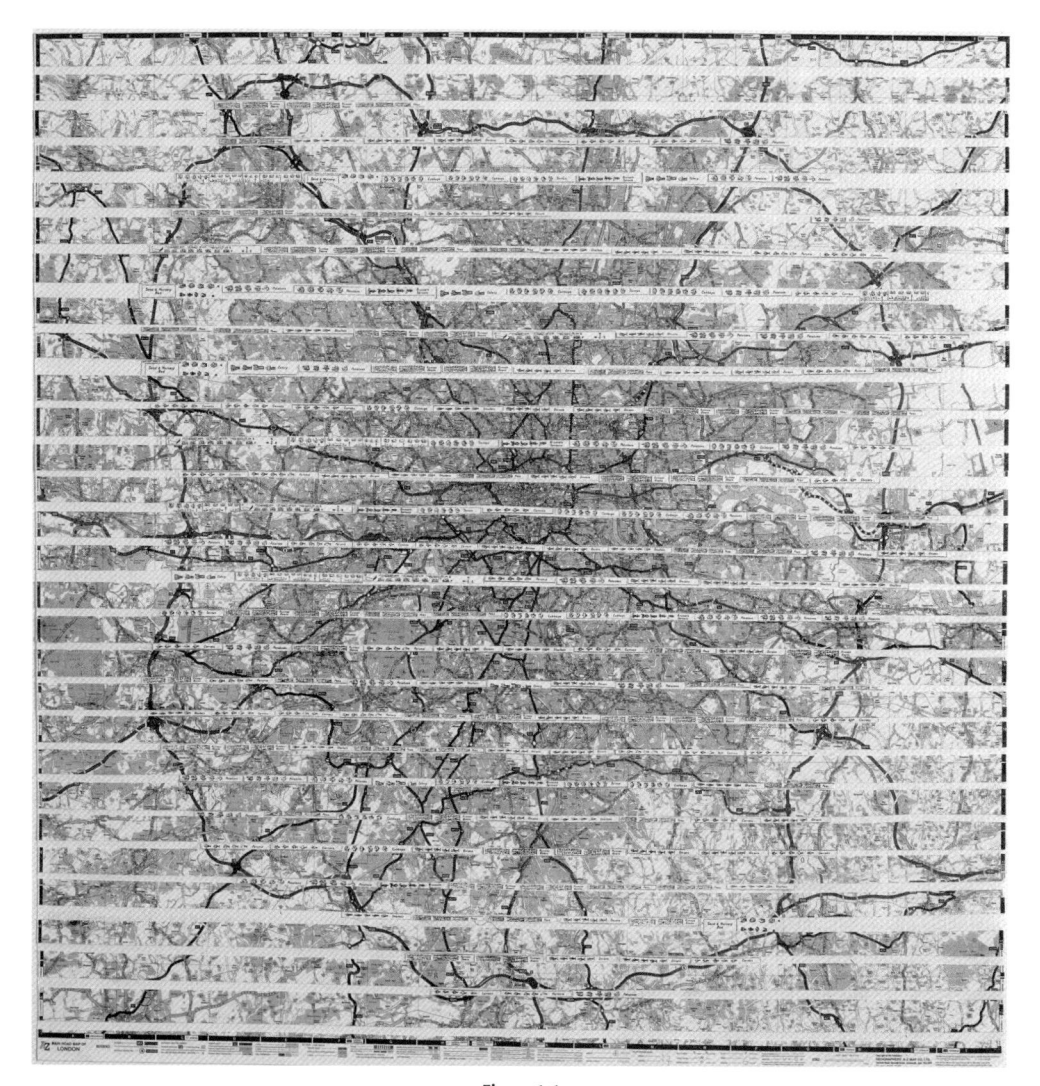

Figure 1.1

Figure 1.1 *London, population 7 million, exploded to accommodate* mini- *and* SUPER-*market gardens. An early study from 1998, demonstrating the area required to supply all of London's fruit and vegetable requirements from urban agriculture. Yields from* mini *and* SUPER-*market gardens are based on 100m² of urban agriculture per person.*

Around midday, people pack their stuff and leave, provided they are not staying for lunch in one of the restaurants or going for a swim in the local open-air pool, or starting work in the work areas. Later, they might walk or cycle to Tate Modern or Covent Garden, out into the countryside or to the River Thames. With a quarter of roads converted to accommodate CPULs since 2005, one can now reach virtually any point in London by walking or cycling in less than an hour or two. Most London boroughs pride themselves on having designed their open space so as to allow people to access a CPUL after no more than a 15-minute walk.

If people don't feel like walking back from where they are now, they can take one of the regular (every 10 minutes) buses or trains, pick up a taxi or hire one of the cars or bicycles which they can later leave close to their home to be used by the next person.

Meanwhile, the CPUL is heaving with children and youngsters engaging in all sorts of sports and fun, with people (and dogs) going for walks, sunbathing or enjoying tea, games and books.

Families gather in the various state-of-the-art activity grounds, which during the week are mostly booked by schools and clubs. The small open-air swimming pool has its busiest time.

In the early evening, the CPUL fills with people coming home or going out for dinner and/or into town. Teenagers meet their friends in the more hidden areas between the fields, musicians start playing and people start to dance. Children do their last races around the lanes and lawns. People have picnics and barbecues, do sports or hire a deck chair to relax. At the same time, others cycle or walk back from work, enjoying the evening air and sun and some quiet activity which will change into urban bustle and business the closer one gets to the city centre or the borough sub-centres.

This evening, the CPUL hosts a film event: a big screen is temporarily hang over its main space, people sit and lie around watching, the restaurants are busy, the pubs and cafes . . .

On a rainy winter weekday, though, things look different. Cyclists pass in rain clothes, people hasten to work, train or car sharing stations. Farmers, who at that time of year prepare soil, seeds and tools for the next spring, work in the CPUL sheds and product stores. The poly tunnels that are now occupying most of the fields open up automatically to soft rain allowing the seasonal vegetables to catch it. A few children play in a playground and explore the demonstration rain water mills – differently sized sculptures that collect rain. Apart from the aforementioned, and some dogs or lovers who enjoy running around in that weather, the CPULs lie empty, sucking in the rain. It is a busy time for delivery services, the CPUL is full of delivery cycles racing to bring food and other shopping to their clients. A man is moaning that his car sharing station has run out of cars, but cheers up when the gas-driven bus, with a five minute frequency, arrives. The weather forecaster celebrates the rain and tries to predict the benefit it will bring to the productivity of particular urban areas. Today, it is also windy but it blows too hard for kite flying or kite sailing, which people would otherwise do. The CPUL is really deserted. The staff on the farmers' market hide under retractable canopies, unless the whole market has already retreated to the small covered or heated market hall, of which there is at least one close to every CPUL. At this time of year, the business booms for imported fruit and vegetables.

Then the rain stops. Now, the refreshed oxygen-rich air, that the wind brought in from the sea via the many open urban corridors, stands clear above the CPUL and its adjacent buildings. Some farmers walk along checking the huge underground rainwater tanks for how well they have filled. Sun will later operate the PV driven pumps that distribute the rainwater to particular fields and houses for productive and private use. The numerous small overground water canals are busy and bustling; children with wellies (and no bellies) sail boats and sticks.

This winter day, indoor activities, both at home and in leisure centres, offer endless possibilities to play, learn, do sports, get involved in the arts, meet up, etc., at any time after work or during the day. The local leisure centres, situated at the CPUL edge, can now always boast outside parts such as swimming pools, racing tracks or sauna seating areas where sitting in the rain is ever so exciting. And for the evening, the city centre is still only a quick train ride away . . .

ECOLOGICAL INTENSIFICATION

The urban strategy that enabled CPULs to happen and to grow to what they are now, in 2045, was called Ecological Intensification (named 'Carrot City' by London architects a few years later). In London (as in most European cities), this incremental strategy has been applied since about the year 2005.

Ecological Intensification worked by prioritising environmental urban layers, which were either connected to open space use, or to implementing sustainable technology and activity patterns. Usually, these environmental layers were then superimposed with other locally appropriate layers, such as economic, social, cultural, historical, etc.

As a result of this process, London has become (and is still becoming) a real 'Carrot City': sustainable, integral, working out of itself, but within its capacities, allowing and needing the participation of its citizens as well as offering real lifestyle choices.

The London boroughs, for example, applied individual development strategies centred around innovative ways of connecting and reconnecting urban work, trade and leisure activities. These were thought of as the basis for an exchange of products and services, and succeeded in supporting local productivity patterns and thereby an economic prosperity all over London.

Over the last 40 years, since about 2005, London has developed into a city where national and international exchange brought mainly those goods and services into the city which it was not possible to provide from within. Through concentration on its local expertise and workforce, London (and Greater London) has grown to strong economic identity through its excellent and eccentric products. At the same time, it regained a national and international market lost for decades because of the closure or takeover of most of its unique factories, farms, foods and fashions. Employment figures in London (and all over Britain) have rocketed over the past twenty years. London was also able to share in tackling the international problems originating from export-based production in developing and poorer countries, such as exploitation, cheap/child labour, mono-cultures or -industries, and uncontrolled environmental damage. Most of those problems had long been recognised as reasons for the large population influx into twentieth century cities in the first place.

All this development has highlighted the exotic in international products, i.e. *food products* such as – example for this book – fruit and vegetables. Contrary to food trading methods in the early twenty-first century, exotic fruit and vegetables were now only sold when ripe and tasty. Suddenly, these 'special foods' could promote their true flavour, colour or texture, celebrating their geographical and cultural difference at London tables. As people now favoured, for various reasons, a majority of locally produced foods, the exotic compliment became again exotic, leaving room to explore the local. Food and eating – an informed choice between staple and healthy, fresh and exotic, local and organic – grew to real importance in people's daily routines. In London (and anywhere else), this did not only result in people's improved general health. Economically, it led to higher payment for better quality farming and retail, higher employment rates due to more careful handling of food, i.e. in smaller retail units, and less food and therefore energy waste, both on a national and international level. Altogether, the changes to the food sector resulted in decreasing environmental problems as over-production, mono-cultures and mass-transportation became issues of the past . . .

Such *shifts in urban lifestyle* were crucial for any *shifts in urban landscape*. In London, as well as other 'Carrot Cities', Ecological Intensification generated not only employment, capital and liveliness, but also a different relationship to open urban space. The new lifestyle options on offer in combination with the multitude of efforts to improve the urban fabric, restored positive attitudes towards the city, discouraging, for example, people from moving out of cities into suburbia, one of the major urban problems of the turn of the century.

LONDON IN 2045: POSTSCRIPT

It is now widely acknowledged that CPULs have grown alongside three main urban prerequisites: population stability, successful public transport and borough balance.

London's population had stabilised since about 2040 at nearly 9 million and both the city's skyline and the city's outline had ceased expanding.

This was mainly due to the reduced influx of people into London. Worldwide, there are now considerably fewer social, economic and political inequalities between countries so that *moving for better life conditions* has been replaced by *moving for richer life experience* which happens fairly evenly all over the globe. As stability does not mean zero-motion, London's changing demographics are a constant source for cultural cross-fertilisation that is most visible in London's vibrant diversity.

Another contribution to London's stable population was that the trend to live single lives in single flats, as observed at the end of the twentieth century and predicted to increase during the twenty-first, had stopped around 20 years ago. Of the many reasons for this change, the most influential one has been a rediscovery of particular lifestyles, with increased numbers of people enjoying, for example, partner(s) or family. The accompanying new work and leisure activities have led to massive pilgrimages to urban amenities, revitalising London's public spaces and its economic prosperity beyond expectations.

Thirdly, London also ceased expanding as a result of better use and management of space within the city. During the past 40 years, this allowed an increase in urban density of 20 per cent with a simultaneous increase in the amount of open urban space. Compared to the year 2001, London's city boundaries now enclose 10 per cent more open space while holding two million more people.

This last measure was at the same time extremely important for the solution of London's two other major problems at the turn of the last century – traffic congestion and borough imbalance – which have since then been constantly reduced.

Apart from the previously described changes to peoples' opportunities for moving through the city, road traffic had been targeted with various measures to reduce the use of private transport. This led mainly to the establishment of state-of-the-art affordable public transport and the rediscovery of the *city of short ways*, i.e. integration of work/trade space within living spaces. The introduction of

CPULs played an important role as it enabled people to choose from and effectively use individual options ranging from walking via cycling, cycle taxis and delivery services to car sharing systems and network buses. Consequently, traffic congestion with its former huge impact on air and noise pollution, low road quality, high road accident numbers, stress, natural resource depletion, etc., has not been considered problematic since about 2030.

Borough imbalance, with its resultant modern slums, suburban sprawl, unequal provision/loss of open space, congestion, crime, quality differences in built developments, etc., has lost its grimness, though it is still an issue.

Over the last 40 years, the equitable development of London boroughs was supported by most public and private bodies through targeted networking in and between the boroughs. A beneficial borough balancing plan in spatial terms was the 'green lung project', which soon became part of the CPUL movement. It invited every borough to participate in the creation of quality local open spaces that were then connected to a regional-urban landscape concept.

2

MORE SPACE WITH LESS SPACE: AN URBAN DESIGN STRATEGY

Katrin Bohn and André Viljoen

WHAT ARE CPULs?

Overlaying the sustainable concept of *Productive Urban Landscapes* with the spatial concept of *Continuous Landscapes* proposes a new urban design strategy which would change the appearance of contemporary cities towards an unprecedented naturalism. Continuous Productive Urban Landscapes (CPULs) will be open landscapes productive in economical *and* sociological *and* environmental terms. They will be placed within an urban-scale landscape concept offering the host city a variety of lifestyle advantages and few, if any, unsustainable drawbacks.

CPULs will be city-traversing open spaces running continuously through the built urban environment, thereby connecting all kinds of existing inner-city open spaces and relating, finally, to the surrounding rural area. Vegetation, air, *the horizon*, as well as people, will be able to flow into the city and out of it. Partially, the city will become open and wild.

CPULs will be green, natural and topographical (except when they happen on buildings), low, slow and socially active, tactile, seasonal and healthy. They will be well-connected walking landscapes. Depending on their individual settings and the urban fragment used, CPULs will read as parks or urban forests, green lungs or wilderness, axes of movement and journey, or places for reflection, cultural gathering and social play. They will be containers for an assembly of various activities that do not happen in buildings.

CPULs will not be about knocking cities down or erasing urban tissue; they do not seek a *tabula rasa* from which to grow. Instead, they will build on and over characteristics inherent to the city by overlaying and interweaving a multi-user landscape strategy to present and newly reclaimed open space. Very importantly, they will exist alongside a wide range of open urban space types, complimenting their designation and design and adding a new sustainable component to the city (see Chapter 14). CPULs will adapt to the various ways in which individual cities develop by tailoring their type and layout to specific urban conditions and fulfilling their own requirements in a loose and inventive manner. Every CPUL and every fragment within it will build up their own individual, constantly changing character.

CPULs will be productive in various ways, offering space for leisure and recreational activities, access routes, urban green lungs, etc. But most uniquely, they will be productive by providing open space for urban agriculture, for the *inner-urban and peri-urban growing of food*. The urban land itself, as well as the activity happening on it, will become productive: occupants will act and produce *on* the ground and *with* the ground. Vegetation will appear ever new and exciting: it will get harvested, grow back, get harvested again, grow again, grow differently, grow less or more, grow earlier, later, it will seasonally change size, colour, texture and smell... Whilst there are various examples in contemporary urban design of establishing green links or open space similar to continuous landscapes, the aspect of agricultural production, of the rural, will add not only an important new spatial quality to the city, but also socio-economic and environmental qualities (see Chapter 3).

CPULs will be designed primarily for pedestrians, bicycles, engine-less and emergency vehicles, so as to allow healthy vegetation and varied occupation. The resulting near absence of noise, air and ground pollution, and of the dangers from traffic, i.e. accidents, would make CPULs not only most appropriate for agricultural production, but also a perfect leisure destination for the local population. Distances and dimensions within the city will change dramatically. With regard to the present condition of European cities, to their congestion, commuter lifestyles and environmental damage, CPULs will be as revolutionary for any of them as the introduction of the underground was for London.

CPULs do not exist yet. However, several types of urban agriculture do already, and will always, exist: city farms, market gardens, allotments, back gardens, community gardens, etc., play already established roles in urban life all over Europe (see Chapter 11). Such structures would form individual components of the new CPULs, with the advantage of open islands now being connected to a widely accessible regional landscape, making them urbanistically more meaningful.

Apart from this connectivity, the main design implication of CPULs will be the introduction of *agricultural fields* into the contemporary city. As the ground – *the earth* – air and vegetation become vitally important for the productive success of CPULs, the effort to maintain their well-being ensures that natural conditions will be most significant features within the new urban landscape. Other characteristics of CPULs will evolve in accordance with the landscape's ecological aims. This is a feature of the rural land.

Depending on their size and location, the spatial types of productive urban landscapes will range from small uni-crop to large multi-crop fields being placed within (and occasionally outside) a CPUL (see Chapter 24).

Generally, any open urban space, communal or private, inner-city or sub-urban, small or big, would benefit from integration into a CPUL. Even fully laid out open urban spaces, i.e. parks, could allocate parts of their land for productive use, gaining in return access and connection to a continuous landscape design, thereby becoming wider, wilder and healthier.

The new design strategy will allow high diversity, as it will benefit from difference and a new identity to enrich the occupation and appearance of its various productive and connective landscape elements. Whilst existing quality open spaces might want to keep their identities and be largely used as before, recycled open spaces will adopt individual design strategies with reference to the planned co-existence of food production, current uses and history of the place.

WHY CPULs?

Continuous Productive Urban Landscapes are about *urban* food growing and *local* consumption. They will include livestock, but consist largely of vegetation which is locally managed: mainly organic vegetables, fruit and trees, planted in rows, planted in groups, fields, patches, etc. Vegetation will be chosen for its inherent extractable energy (i.e. it can be eaten) or its material quality (i.e. it can be worn), then grown, harvested, traded and consumed. The main production will be carried out by local occupants who rent the land and work it commercially within an individually defined local framework (see Chapters 17 and 25). Cities that decide to support this concept of organic local farming, trading and seasonal consumption will never be fully self-sufficient in food production. They will still be required to bring food into the city from the hinterland, but less of it and in a more focused, need-oriented way.

Producing food where one wants to eat it, or consuming food where it has just grown, establishes a healthy and sustainable balance of production and consumption. It is an effective and practical, but at the same time self-beneficial way of reducing the energy embodied in contemporary Western food production.

This reduction of embodied energy is crucial for several reasons. The energy – mainly non-renewable – currently used for conventional food

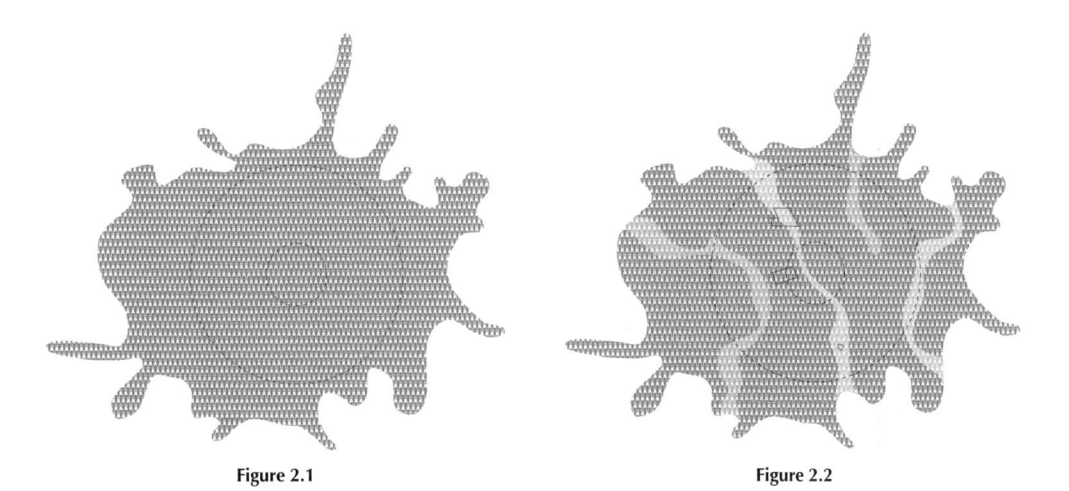

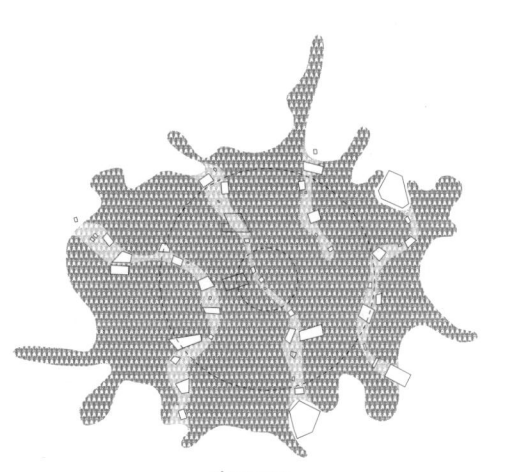

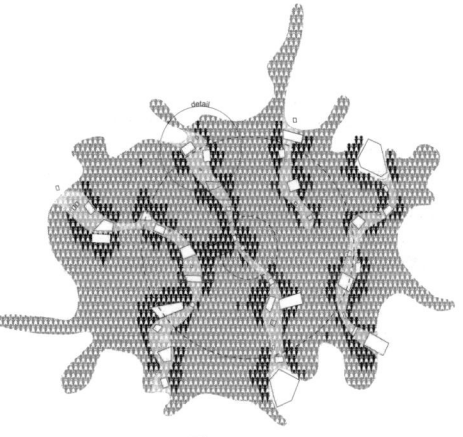

MORE SPACE WITH LESS SPACE: AN URBAN DESIGN STRATEGY

Figure 2.1

Figure 2.2

Figure 2.3

Figure 2.4

Figure 2.1   *An established city with no CPULs.*
Figure 2.2   *Identifying continuous landscapes.*
Figure 2.3   *Inserting productive urban landscapes.*
Figure 2.4   *Feeding the city.*

CARROT AND CITY: THE CONCEPT OF CPULs

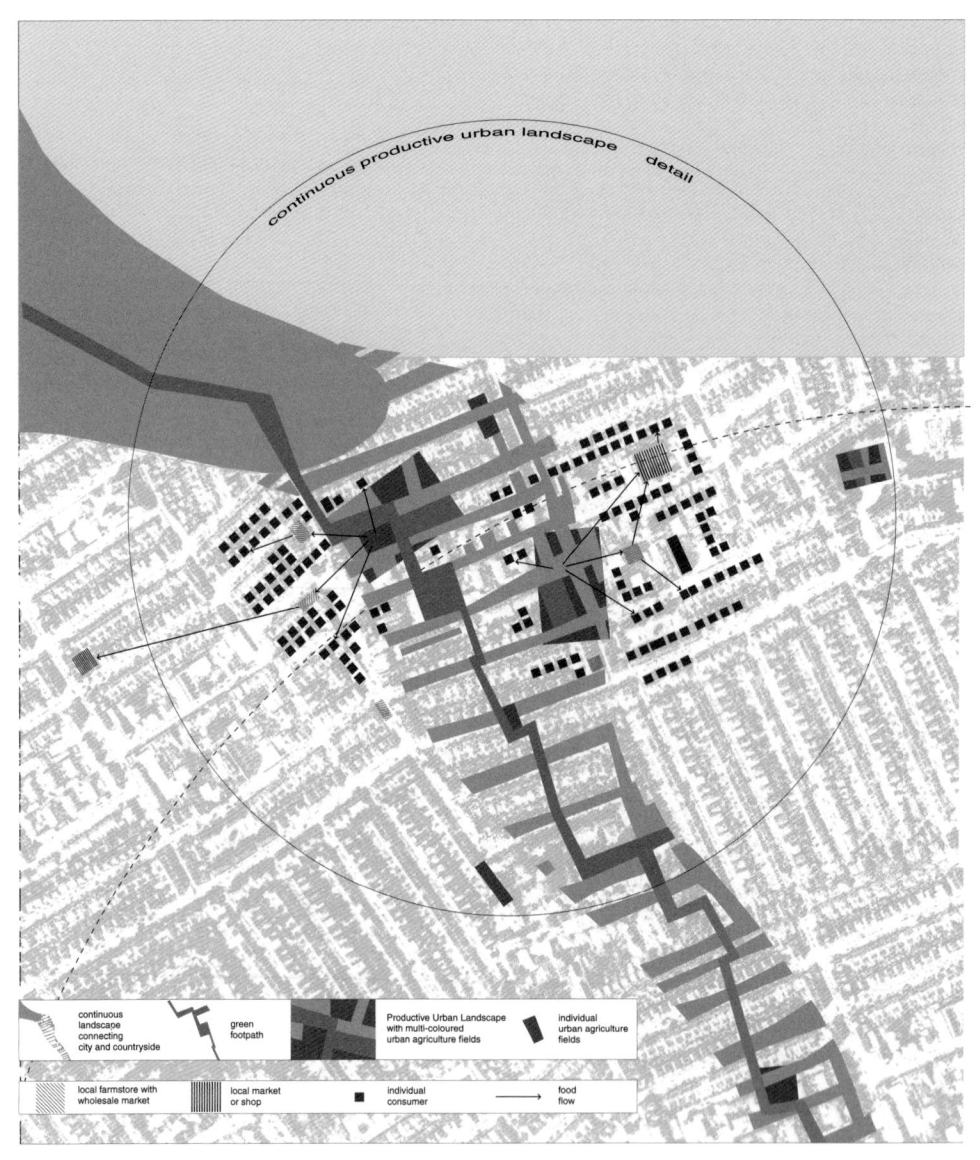

Figure 2.5

production, for example in Europe, exceeds by far the energy received in return from consuming the produced food. The unlimited, daily usage of non-renewable energy contributes significantly to global resource depletion and, through greenhouse gas emissions, to global warming (see Chapter 3).

One might argue that concentrating on the energy consumed by current remote food production does not allow for the future development of environmentally clean energy technologies. But, firstly, such a position fails to recognise that the inequitable distribution and consumption of resources extends beyond energy usage, i.e. to raw materials, desirable land, water and food. Reducing the energy requirements of goods and processes shrinks the divide between those who have access to abundant energy supplies and those who do not, without limiting the availability of final products.

Secondly, CPULs will not solely concentrate on addressing embodied energy in the first place. They will be *environmentally productive* dealing not only with local food, but also with issues such as greenhouse gas (CO_2) reduction, improving air quality and air humidity, noise filtering and biodiversity. They will also be *sociologically productive*: their urban concept will involve, amongst others, cultural, educational and leisure activities, shopping habits or diet and health concerns. Last but not least, they will be *economically productive*, provoking new strategic socio-economic thinking and changing local employment and product–cash flow patterns (see Chapter 8). CPULs will provide a completely different concept for the use of the city.

WHERE WILL CPULs BE?

CPULs require land, but in return they will enrich cities by reducing their environmental impact and bringing in spatial qualities until then only associated with rural or natural conditions.

The implementation of CPULs will be a slow process varying with the city under consideration. Each city will have to determine the scale and ambition for CPUL infrastructures. A long-established large city, like London for example, provides less scope for integrating CPULs and/or tracts of productive urban landscape close to its historic centre than it does close to its fringes (see Chapter 24).

Above all, the new open urban space will need . . . space. Land will have to be allocated, reclaimed, recycled or imaginatively found. Open urban space is precious: it is rare, and it can be converted into something else. Usually, it is converted into housing or other building development and thereby into money. CPULs will have to compete with commercial land-use activities, as well as with all other contemporary ways of designing or redesigning open urban space (see Chapter 14).

Urban agriculture, the proposed productive element of CPULs, could take on any shape and occupy virtually any space in the city – big, small, horizontal, sloped, vertical, rectangular, triangular, irregular, on brownfield sites, on greenfield sites, in parks, on reclaimed roads, on spacious planes or squeezed in corners . . . *CPULs will appear in ever varying site-specific shapes and dimensions and at any urban scale*. They could happen anywhere within the urban context, leading to many cities boosting the multiple use of their build space *and* keeping valuable inner-city space clear of construction at the same time.

Starting to implement CPULs will be most advantageous in those areas of the contemporary city which, at the moment, are given over to activities restricting space to mono- or non-use: car parks, roads, parking bays, shopping malls, warehouses, multi-storey car parks (at least the flat roofs and plane facades), railway embankments, industrial wasteland, brownfield sites, etc. These inner-urban spaces are available in abundance with occupants and users being mostly positive about qualitative spatial and environmental improvement. Moreover, these spaces' diversity in size and shape, and their location anywhere in urban networks make them ideal components of the continuous landscape strategy.

The emphasis of CPULs on connection and movement, on uninterrupted routes between local urban centres, will influence their layout and positioning (see Chapter 25). Whilst they will hold areas sufficiently sized for many activities, there will be other parts which will mainly provide continuous and direct urban connections. Both of these CPUL types will draw from well-established, existing or newly reclaimed open urban space. An extensive network of CPULs could be imagined as a contemporary city with every other road given over to farming, leisure, and pedestrian and cycle routes (plus occasional and emergency vehicular traffic), thereby connecting larger existing and new open spaces of any size and designation. It could look like Venice with the canals transformed to become fields.

However, reclaiming and recycling those areas poses two problems: the proposed sites could be toxic and their current mono-use activities would have to be accommodated somewhere else, either by redesign or relocation. Successful attempts to do this have taken place in various countries, mainly in connection with the necessary repair of existing urban tissue.

As costly and time-intensive as such solutions might be, and as much focused urban and architectural planning or communal support they might need, they will always have the advantage of being environmentally sensible *in addition* to providing the urban and socio-cultural benefits described above (see Chapters 3 and 17).

PART
TWO

PLANNING FOR CPULs:
URBAN AGRICULTURE

3

MORE FOOD WITH LESS SPACE: WHY BOTHER?

André Viljoen, Katrin Bohn and Joe Howe

By agriculture only can commerce be perpetuated; and by Agriculture alone can we live in plenty without intercourse with other nations. This therefore is the great art, which every Government ought to protect, every proprietor to practice, and every inquirer into nature improve.

Dr Samuel Johnson 1709–1784 (Johnson, 1756)

By the year 2025, 83 per cent of the expected global population will be living in developing countries. . . . Agriculture has to meet this challenge. . . . Major adjustments are needed in agriculture, environmental and macro-economic policy, at both national and international levels, in developed as well as developing countries, to create conditions for sustainable agriculture and rural development.

Agenda 21 (United Nations Conference on Environment and Development, 1992)

It is self-evident that food production is fundamental to life and that it underpins all other activities. The Earth summit of 1992 recognised this and the need for adjustments in the current means of agricultural production.

Dr Johnson, writing in eighteenth century England, notes the role agriculture can play in supporting commerce, in this instance by making a nation self-sufficient. While we do not advocate a world 'without intercourse' between nations, a strong case can be made for modifying current practice in globalised food production and distribution by increasing local self-sufficiency in food production. An example of the impact globalised food markets can have on local economies is provided by a press release from the international anti poverty charity ActionAid:

More than 75 per cent of Kenya's population depends on agriculture, but subsidised imports are destroying the country's market.

ActionAid Kenya subsidies expert Gichinga Ndirangu says farmers cannot compete against low-priced, subsidised products, such as wheat, which are flooding into the country. 'Because farming in Kenya is no longer subsidised, the cost of production is much higher,' he says. 'Kenyan farmers can not compete.'

During recent years they have been forced to reduce their production of wheat because they can no longer break even. They've looked for alternative crops, but with the entire agricultural sector facing similar challenges, it has been difficult to find viable substitutes. 'Kenya is one of the countries that, in the mid-80s, was compelled to cut support to their domestic sector and open up their markets to cheap foreign imports as conditions of loans from the World Bank,' explains Gichinga. 'The Bank argued these measures would allow people access to cheaper products and alleviate poverty – of course it has had the opposite effect. All Kenyan farmers want is to be able to make a living from agriculture. And the only way to do this is by protecting them from cheap subsidised imports.'

The second reform was to focus on growing commercial crops for export and then rely on foreign exchange earnings to pay for importing the food it needed.

Today Kenya is no longer a self-sufficient producer of basic foodstuffs. The country has become increasingly vulnerable to the vagaries of the world market. Kenya must now pay for vital food imports with the money it earns from exports.

Unfortunately, the prices of two of its major crops – tea and coffee – are currently falling because of a glut on the world market.

(ActionAid, 2002)

FOOD AND URBAN DESIGN

Let's start by considering how we have reached this state of affairs and why architects, urbanists and planners have a role to play in improving the situation.

In the English speaking world, the publication of Carson's *Silent Spring* in 1962 triggered concern over the ecological side effects and health risks posed by industrial agriculture and agribusiness, resulting in chemically dependent farming techniques. Today, many rural areas have been reduced to biologically impoverished wastelands and in Britain, for example, the domination of industrial agriculture is creating an increasingly depopulated landscape.

The increased disconnection between consumers and producers of food means that urban populations have little connection with food production and thus have a limited knowledge of the issues associated with it.

This process makes the world less comprehensible and reduces the ability of populations to criticise the status quo, due to a lack of direct experience and knowledge. Nevertheless in Europe there is much public concern about industrial agriculture, for example the application of genetic engineering to food production. Research by the European Commission has indicated that the majority of consumers in Europe would be prepared to pay a premium for non-genetically modified food (European Commission, 1998).

A number of recent food scandals in Europe, such as Mad Cow disease in the UK, the contamination of Austrian wine, and dioxins found in Belgian eggs and meat, have resulted in health concerns about industrial agriculture. Some argue that the public is concerned that regulations are not enforced rigorously enough (Aerni, 2000), but we think these concerns indicate a wider scepticism regarding the overall benefits of highly competitive and market driven agriculture managed by a very small group of businesses. In support of this counter argument, Aerni refers to reports noting that Japanese consumers are ready to pay high subsidies to maintain family-based multifunctional farming and that they use protectionist policies to ensure this (Hayami and Godo, 1995).

WHY URBAN FOOD?

Urban agriculture can result in environmental, social and economic benefits. There are three primary environmental benefits from organic urban agriculture – preserving biodiversity, tackling waste and reducing the amount of energy used to produce and distribute food.

Modern industrial farming techniques in the countryside have had a devastating effect on biodiversity. The combination of fertiliser and pesticide use with habitat destruction means that urban environments are now often more species-rich in fauna and flora than their rural counterparts (Nicholson-Lord, 1987).

Added to this has been the effect of a few large supermarket chains dominating food retailing. For example the United Nations Food and Agriculture Organisation cite Belgium, France and the United Kingdom as 'extreme examples' where only 10 per cent of retail units account for more than 80 per cent of food distribution (Food and Agriculture Organisation, 2002). Supermarkets' reliance on economies of scale and repeatable quality standards inevitably favours larger suppliers and the use of chemicals in preference to environmentally benign agricultural methods. In contrast, urban food production, particularly current forms such as urban farms and community gardens, tends to be characterised by the use of organic methods and the local sale of produce.

Urban agriculture also offers the potential to use organic waste for composting, thereby reducing the need for land-fill (see Chapter 12).

Food is being transported further than ever before, often by air between countries on opposite sides of the world, whilst local crop varieties are replaced by a few commercial types popular with supermarkets (Cook and Rodgers, 1996). This pattern of growing 'food miles' is far from sustainable, its by-product being increasing air pollution, notably of major greenhouse gases such as carbon dioxide, increasing road congestion and noise, and increasing stress. Urban food production supplying local outlets offers an alternative to this pattern (see Chapter 5).

The consequences of this are evident in the reduced numbers of varieties of particular fruit and vegetables available in most supermarkets; reactions against this include the Italian 'Slow food movement' which promotes the use of fresh local produce and the associated culture of convivial eating.

At the same time, many people in poorer urban parts of European cities are living in areas that are effectively becoming retail deserts (see Chapter 6). Such factors underline the unsustainability of current trends in food retailing and production. Despite this, the whole issue of food security and food supply with its attendant environmental, social, economic and health knock-on effects is one which very few municipal or national authorities have addressed.

THE ENVIRONMENTAL CASE FOR URBAN AGRICULTURE

One of the most effective ways of assessing the environmental impact of a particular process or product is to find out how much non-renewable energy is required to produce it; this quantity of energy is referred to as embodied energy. The consumption of embodied energy results in the emission of greenhouse gases, which contribute to global warming and climate change. So embodied energy can be thought of as a shorthand for assessing the climate change potential of a process.

Another important reason for finding out how much energy different processes and products use is to judge how equitably the world's resources are distributed. In 1985 the per capita carbon dioxide emissions for someone living in Africa was 0.79 tonnes per annum while the corresponding figure for someone living in the USA or Canada was 19.21 tonnes per annum (Shorrock and Henderson, 1989). The challenge lies in reconciling the need for reducing emissions of greenhouse gases and the imbalance in access to energy between developed and developing regions. Most people would prefer this imbalance to be evened out by a general improvement in the living standards of those with the least access to resources. A variety of mechanisms can achieve this end, all requiring access to energy, and if you think, as we do, that conventional nuclear power is not an attractive and safe long-term solution, then the only option is to use renewable energy and reduce the amount of energy required for processing goods and activities. The need to reduce embodied energy applies as much to food production as to the energy efficient operation of buildings and other activities.

Figure 3.1 was published by Peter Chapman in 1975 and illustrates the embodied energy for a loaf of bread (Chapman, 1975). This is typical of a number of studies conducted in the 1970s, undertaken in response to the rapid increase in the price of oil, and fear over the impact this and possible oil shortages, would have on the economies of developed countries. Between then and the Rio Earth Summit in 1992, very little attention was paid to embodied energy.

In Britain, the first widely published attempt to assess the embodied energy of food was made by the architects Brenda and Robert Vale (Vale and Vale, 2000). They tried to estimate the non-renewable energy use resulting from food consumption for a family of four, living in England. This was compared to the family's energy usage in their house and for the typical usage of a family car. This study took as a starting point the recommendations for daily calorie intake by the Food and Agriculture Organisation of the United Nations. The energy content of food was estimated using United Kingdom data from 1968 published by G. Leach (Leach, 1976). It is clear that the figures used in this calculation are out of date and that the embodied energy of food may have altered. For example, packaging and transportation have almost certainly increased, while production and processing may have become more energy efficient. Notwithstanding these uncertainties, the Vales' findings are sufficiently surprising to cause concern.

They estimated that the embodied energy content of food consumed by the family of four was 265 kWh/m^2, while the delivered energy used by a typical house in the United Kingdom was estimated at 257 kWh/m^2. In both cases, to allow for comparison between houses of different sizes, the total energy used per year was divided by the floor area of the house.

Figure 3.2 shows the results of extending this calculation to take account of the associated carbon dioxide emissions. Here, the Vales have estimated that the carbon dioxide emissions related to food production may well equal the amount of carbon dioxide emitted from the combined use of a private

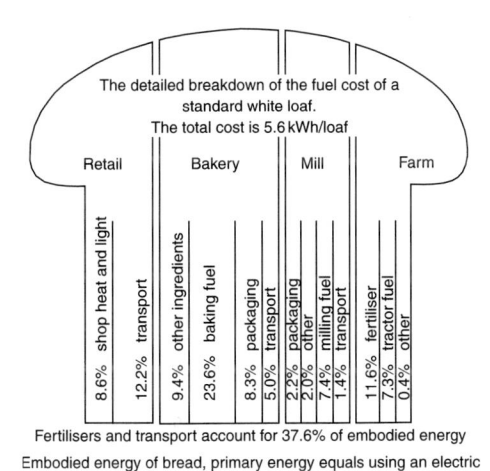

Figure 3.1 *The embodied energy of a loaf of bread.*

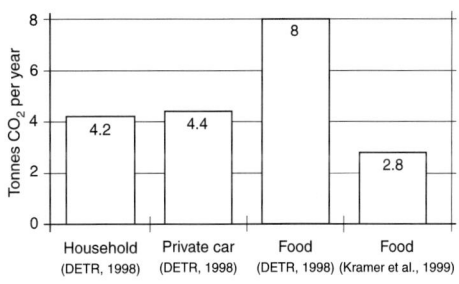

Figure 3.2 *Carbon dioxide emissions for a typical family in the UK.*

car and a standard house built in England in 1995. Figure 3.2 also includes estimates from a Dutch study for household greenhouse gas emissions resulting from food consumption (Kramer K.J. et al., 1999).

The study by Kramer et al. was detailed and used contemporary figures to estimate emissions resulting from the lifecycle for food production and consumption. Kramer et al. estimates that the food consumed per year by an average household resulted in 1875 kg of carbon dioxide emissions. When other non carbon dioxide greenhouse gas emissions resulting from food production are allowed for, then annual household food consumption emits the equivalent of 2800 kg of carbon dioxide. A carbon dioxide equivalent refers to the amount of carbon dioxide that would have to be emitted to equal the global warming potential of another greenhouse gas. In the case of food this refers to the greenhouse gases methane and nitrous oxide.

Kramer et al.'s study indicates that the global warming potential resulting from the consumption of industrialised food production in the Netherlands is about equal to the emissions for all energy use in a typical modern home. For comparison, circa 2600 kg of carbon dioxide emissions result from space heating, domestic hot water use, cooking and the use of pumps and fans in a typical four person house built in the UK to thermal standards applicable between 1995 and 2002 (DETR, 1998).

The Dutch study suggests that carbon dioxide equivalent emissions are about 40 per cent of those estimated by the Vales. Given the number of variations between the two calculations it is difficult to make a direct comparison. The Vales' calculation is based on a household of four persons which is probably larger than the national average household in the Netherlands, and the Vales' embodied energy data is much older.

Another way of assessing the relative importance of greenhouse gas emissions from food production is to compare them to emissions from the embodied energy of buildings. Figure 3.3 assumes a house with a floor area of 100 m^2 for four persons. From Figure 3.3, using Kramer et al.'s figures for carbon dioxide emissions due to food consumption, it can be seen that the embodied energy of food is significantly greater than the embodied energy of dwellings. (The embodied energy of dwellings has been calculated as an annual equivalent figure, by dividing the embodied energy of building materials by their lifetime.)

Overall these estimates for the embodied energy and associated greenhouse gas emissions of food are startling, particularly since the energy usage associated with agriculture is usually considered to be relatively small. Europe provides a useful illustration since its per capita energy use falls between that of the highest consumers, the USA and Canada, and the lowest, Africa (Shorrock and Henderson, 1989). The United Nations Framework Convention on Climate Change states that between 1990 and 1997 agriculture contributed less than 1 per cent to total carbon dioxide emissions, rising in 1998 to 1 per cent (European Environment Agency, 2002).

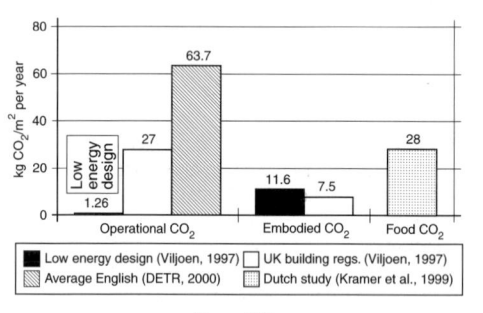

Figure 3.3

Figure 3.3 *Embodied and operational carbon dioxide emissions for a typical household in the UK compared to embodied carbon dioxide emissions from food consumed by a household. The embodied emissions for food are based on a Dutch study.*

MORE FOOD WITH LESS SPACE: WHY BOTHER?

How is it that estimates by some researches claim such a high environmental impact for food production when the carbon dioxide emissions from agriculture appear to be so low?

One reason is that the energy usage on a farm is but one part of the food production cycle. Agricultural energy usage will not take account of the transport required to shift crops from farm to point of sale, and from point of sale to point of consumption. In addition, food in Europe and other developed countries is undergoing a considerable amount of pre sale processing and packaging. Even uncooked fruit and vegetables are frequently sold pre-cut and ready to eat, not to mention the energy used to keep semi-ripe fruit and vegetables refrigerated.

Another significant contributor to greenhouse gas emissions from industrialised agriculture arises from the use of fertilisers. It has been estimated that the energy used in the industrial production of fertilisers and pesticides accounts for 1.5 per cent of the United Kingdom's carbon dioxide emissions. Once fertilisers have been applied to crops further emissions of nitrous oxide occur, a greenhouse gas 310 times as powerful as carbon dioxide. The same estimates suggest these emissions are more than twice as significant as the carbon dioxide arising from the production of fertilisers (Stanley, 2002). One must assume that these significant greenhouse gas emissions are excluded from the agricultural category, and included under emissions from industry.

Using statistics from a range of national institutions, Stanley has suggested that if food in the United Kingdom were produced organically, consumed locally and eaten when in season, greenhouse gas reductions in excess of 40 million tonnes of carbon, per annum, could be achieved. This would represent a reduction of 22 per cent of current carbon dioxide emissions from the United Kingdom. This is put into perspective when it is realised that this reduction is in excess of twice the commitments made by the United Kingdom under the Kyoto Protocol (Stanley, 2002). Similar reductions in greenhouse gas emissions can reasonably be assumed for other nations using industrial agricultural production.

The above study shows how important it is to take account of the entire lifecycle of any product or activity. The principles of organic agriculture, local trading and seasonal consumption of food form one of the central underpinning arguments in support of urban agriculture and productive landscapes. Stanley's timely estimates provide a powerful environmental argument in favour of urban agriculture.

Given the significance of these assertions it is necessary to further investigate the environmental consequences of adopting urban agriculture. Our definition of urban agriculture within continuous productive landscapes assumes the following:

- Organic agriculture
- Seasonal consumption
- Local growing and trading of crops.

ORGANIC URBAN AGRICULTURE

Figure 3.4 indicates how between 1910 and 1970 the energy inputs to food produced in the USA

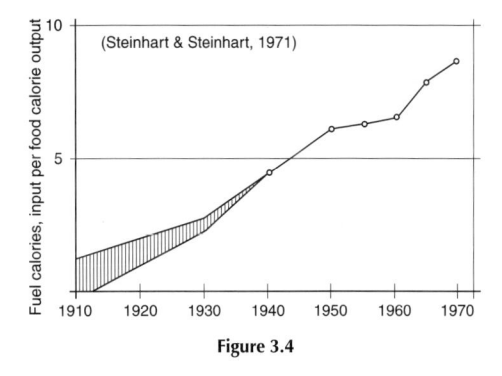

Figure 3.4

Figure 3.4 *Energy input to food production from farmer to consumer in the USA: the number of calories put into the system to obtain one calorie of food. Probable values used for 1910 to 1937.*

increased. Up to about 1920, the amount of energy used to produce and supply food to the consumer roughly equalled the amount of energy released when the food was eaten. By 1970 the amount of energy used to produce food had increased on average by eight times. So for every one unit of energy supplied to a person when eating food, eight times as much energy had gone into producing that food. This is a significant increase in the energy required for food production.

Ratios between the energy content of food when eaten and the energy used to produce the food are referred to as energy ratios. In the United Kingdom detailed energy ratios have been calculated for a number of agricultural products. These took account of all the associated energy inputs up to the point where crops left the farm. In 1968 the average energy ratio for all foods in the United Kingdom was 0.2, which means that for every joule of energy provided by eating food, five joules were used to produce that food (Leach, 1976). Although less than the ratio of eight to one for food in the USA at that time, the trend of increasing energy inputs to food was similar.

When thinking about urban agriculture, we are assuming that during the initial stages of its integration into cities, fruit and vegetables will form the principal produce. This is on the basis that these crops provide the highest yields and value per area cultivated. In order to estimate what impact organic production might have on the embodied energy of crops we undertook a study of Leach's figures for the embodied energy of farm crops (Leach 1976). Although the actual figures used by Leach relate to years between 1968 and 1972, the study gives an indication of potential energy savings. It should be noted that Leach's figures include energy inputs up to the point at which a crop leaves the farm. They take no account of processing, packaging and distribution to point of sale.

Figure 3.5 shows how the calculations were done and Figure 3.6 presents results for a range of crops.

These calculations take account of the possible reduction in yield due to the conversion from conventional agro-chemical-based production to organic farming. In this assessment we have used a worst case assumption – that organic farming yields are two thirds of those from conventional farming (Wright, 1994). This assumption leads to a relative reduction in benefits of organic production. However, some more recent estimates for yields from organic production are less pessimistic. Stanley refers to the Royal Agricultural Society for England, who state that, 'Organic vegetables (yields) are comparable (with conventionally farmed) and potatoes 50 per cent less.' (Stanley, 2002).

Figure 3.7 illustrates the situation usually presented for organic farmers. Yields per hectare may be lower than those for conventional farming, but the price paid per unit output is higher. The relationship between yield and price varies between one crop and another, and so in some instances organic production is more profitable than conventional production and in other cases less profitable (Lampkin and Padel, 1994). If organic yields for fruit and vegetables are similar to those from conventional production, as Stanley notes, then the economics of organic farming become far more attractive. In all of the above cases no allowance has been made for the added value provided by organic farming, for example, the promotion of biodiversity, reductions in greenhouse gas emissions and less leaching of fertilisers into groundwater.

Seasonal consumption

Earlier in the text, the consumption of local seasonal crops was referred to as an important element of urban agriculture. Not only do seasonal

MORE FOOD WITH LESS SPACE: WHY BOTHER?

Organic food production: energy expenditure in terms of energy ratios (after Leach, 1976)

INPUTS	CONVENTIONAL PRODUCTION	ORGANIC PRODUCTION
Fertiliser N, 175 kg	14.00	
Fertiliser P, 175 kg	2.45	
Fertiliser K, 250 kg	2.25	
Field Work, fuels for tractors (to harvest)	2.85	2.85
fuels for harvester, transport	3.38	3.38
Field Work, tractors depreciation and repair	1.14	1.14
harvesters depreciation and repairs	6.70	6.70
Sprays, 13kg	1.24	
Seed shed fuels (620 MJ/t seed)	1.57	1.57
Storage (1,65 kWh/net t)	0.57	0.57
	TOTAL 36.15	TOTAL 16.21
OUTPUTS		
Gross yield t	26.3	
Net yield (less 2.5t seeds) t	23.8	
Edible yield t	17.9	at 66% of conventional 11.9
Energy output (17.9t x 3.18 MJ/kg) GJ/ha	TOTAL 56.9	TOTAL 37.95
Protein output (17.9t x 2.1% protein) kgP/ha	376	
RATIOS		
Energy out/in	1.57	2.34

An energy ratio is defined as the edible energy output of food divided by the energy input necessary to produce it.

Figure 3.5

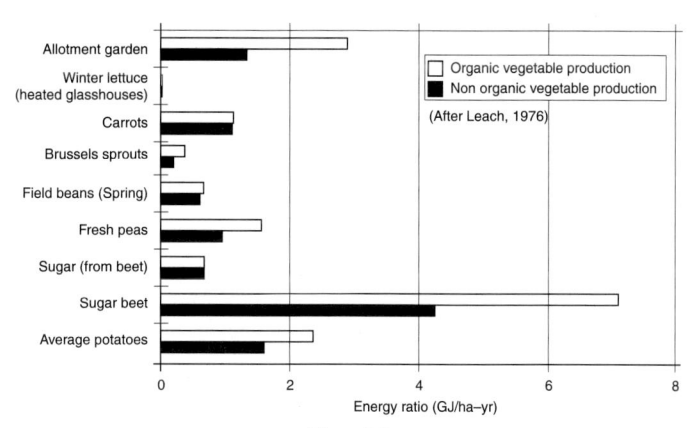

Figure 3.6

Figure 3.5 *Comparison of energy ratios for conventional and organic potato production of UK. Energy input and outputs measured in GJ per hectare per year.*

Figure 3.6 *Comparison of energy ratios for conventional and organic production of UK crops. An energy ratio is defined as the edible energy output of food divided by the energy input necessary to produce it.*

Figure 3.7

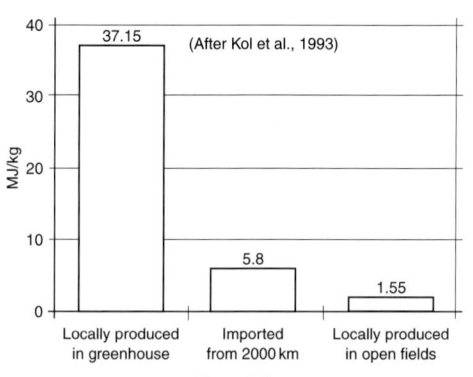

Figure 3.8

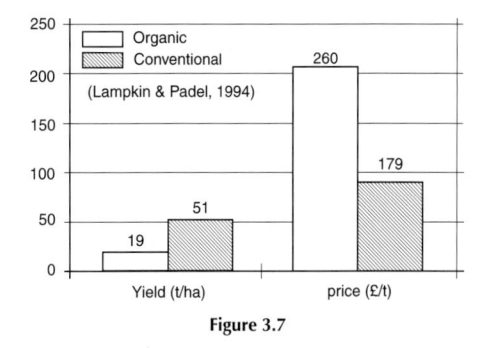

crops reduce the need for the transnational shipment of foods, and hence reduce embodied energy due to transport, but also they have implications for how crops are grown locally.

One way of extending the season of crops is by using greenhouses. These provide a means of growing crops earlier and longer than would be possible in the open. Greenhouses using the sun's energy to warm them provide an effective low energy solution to extending the growing season. But in Europe, for example, the pressures of agribusiness demand ever-earlier crops and consequently many greenhouses are heated, so that yields are increased and the growing season is extended. A study undertaken in the Netherlands has indicated that on average vegetables produced in greenhouses require over 57 times as much non-renewable energy to produce compared to the same vegetables grown in open fields, see Figure 3.8 (Kol, Bieiot and Wilting, 1993).

The desire to consume the same fruit and vegetables all the year round is one of the more important causes of large-scale greenhouse gas emissions associated with food production, resulting as it does in transport requirements and greenhouse heating.

In developed countries, marketing has effectively reduced seasonality to a folk memory. The argument goes that access to all varieties of fruit and vegetables, all year round, provides unlimited choice, and thus the best of all worlds. There are of course merits to this argument, but it does not provide the only answer to a fulfilling culinary life. The obvious drawbacks are those related to international food transport. Within the issue of food transport, environmental matters only worsen as attempts are made to solve inherent problems with the system. We can illustrate this with the situation in England. Currently, many fruits are available in supermarkets that would not be available if grown outside locally. These fruits, when bought, tend to be unripe, because they have been picked early, refrigerated and transported to the supermarket. Experience shows that these fruit never achieve a natural ripeness and rot before they are ready to be eaten. On the other hand they look perfect and by now many consumers cannot compare the quality of these cheap fruits, to quality fresh produce.

One solution to the above problem generated within the food industry is to transport fruit and vegetables

Figure 3.7 *Comparative yields and sale price for conventional and organic crops in the UK, as reported in 1994.*

Figure 3.8 *Comparative embodied energy for vegetables produced in open fields or in heated greenhouses, based on a Dutch study.*

by air, so reducing the time between harvesting and consumption. As Angela Paxton points out in her chapter on food miles (see Chapter 5), this only worsens the environmental impact of the goods.

Local growing and trading of crops

There is no particular reason why fresh, local, seasonal food could not be promoted as powerfully as the limited number of international foods which are available throughout the year. It is understood that urban agriculture will not supply all food needs and that a degree of imported staple and 'special' food will always top up basic food or enliven the pleasure of eating. However, with a well-established local food market, these imports could be kept to a reasonable minimum.

One of the arguments raised against this principle is that developing countries rely on food exports for foreign currency. But as made clear earlier, the prevailing models of trade are not the only options available. The Fair Trade movement demonstrates a viable alternative trading pattern, which supports local production by providing local growers with relatively higher incomes by eliminating a number of unnecessary intermediate traders, etc. Supporting local self-sufficiency internationally would also alleviate problems caused by the unsustainable use of farmland for export crops. As Dr Samuel Johnson noted in 1756 (Johnson, 1756) this would provide a sound base from which nations could develop their economies.

A question arises as to how locally grown food should be distributed within the city. Urban agriculture sites, on the scale of those found in Cuba, circa 1000 to 2000 m² in area, can rely on farm gate sales and local distribution by small vehicles or bicycles with trailers. A radical shift in marketing strategies could result in the use of existing distribution networks of supermarkets. Doing this would integrate urban agriculture into the complex food distribution network of pan national food suppliers. These distribution networks rely on moving goods from the place of production, to processing and packaging plants, to distribution centres from which they are then redelivered to supermarkets for sale. A well publicised study by Stefanie Böge showed that the components for a 150 g pot of strawberry yoghurt will have travelled a total distance of 1005 km before reaching a supermarket shelf in Southern Germany (Böge, 1993). Such complex distribution and production networks rely on integrated large-scale distribution networks.

Local farmers' markets provide an alternative distribution model. At these markets farmers bring their own produce into the city for selling. In order to judge the environmental impact of local farmers' markets in London we undertook a small survey of a newly established farmers' market located in Peckham, about five kilometres from the city centre. This market is one of those set up by Nina Planck (see Chapter 10). We were interested in assessing the transport requirements of small local markets in a large city. Three differences had been noted in the transport associated with local markets, when compared to a supermarket. Supermarkets use large road vehicles to deliver goods from their central storage depots to their retail outlets. This is in addition to the transport associated with collection of goods from farms and any pre sales processing which may occur. We estimated that 90 per cent of supermarket customers drove by car to and from the supermarket.

By contrast, at a local farmers' market each stall holder will drive between their farm and the market. For the fruit and vegetables sold at these markets minimal pre-processing and packaging occurs. It was also noted that nearly all of the market customers walked or cycled to the farmers' market.

Six market holders were surveyed during March 2001. They were asked the round trip distance travelled to attend the market, their fuel consumption, the volume of goods brought to the market, the percentage of stock unsold, the percentage of goods damaged in transit, and the number of markets attended per week. Figure 3.9 presents the results.

We then considered how many people the market could serve. It was estimated that the volume of fruit and vegetables purchased by a family of four would equal about one third of a supermarket trolley per week, which we equated to one tenth of a cubic metre. The average round trip distance travelled by a supplier to the market was 160 km, and allowing for goods unsold and damaged in transit, on average each supplier sold 4.5 cubic metres of goods. Thus on average suppliers travel about 3.6 km per family fed.

As we had observed that most supermarket users drove there by car, and that farmers' market users did not, we concluded that the distance driven by farmers to urban markets would probably be no more than the distance driven by supermarket users. Supermarkets would in addition require the delivery of goods to the store, which as noted involves significant distances. Although this study is small and does not provide conclusive evidence, these results are interesting because they are based on peri urban agriculture. Once urban agriculture is introduced to cities, the distance travelled by farmers will be further reduced.

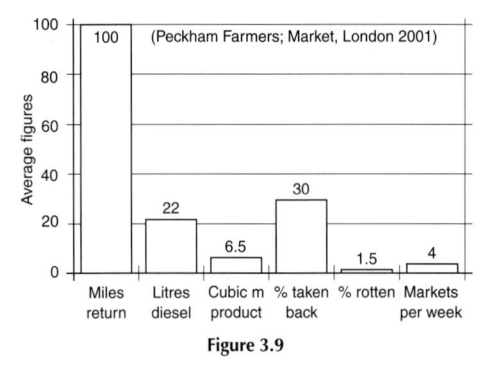

REFERENCES

ActionAid (2002). ActionAid supplement. *Metro newspaper*, 9 September 2002, p.24.

Aerni, P. (2000). *Public Policy Responses to Biotechnology*. STI/CID Policy Discussion Paper No. 4. Harvard University MA. Accessed 19 August 2003 at http://www2.cid.harvard.edu/cidbiotech/dp/discussion9.pdf

Böge, S. (1993). *Road transport of goods and the effects on the spatial environment-condensed version*. Wuppertal Institute.

Chapman, P. (1975). *Fuel's Paradise: Energy Options for Britain*. Penguin Books.

Cook, H. and Rogers, A. (1996). Community food security. *Pesticide Campaigner*, **6 (3)**, 7–11.

DETR (1998). *Building a Sustainable Future: Homes for an Autonomous Community, General Information Report 53*. Department of the Environment, Transport and the Regions/BRECSU, HMSO, Vale, R. and Vale, B.

European Commission Directorate General XII (1998). *Biotechnology: Opinions of Europeans on Modern Biotechnology*. Eurobarometer 46.1 European Commission Brussels/Luxembourg.

European Environment Agency (2002). *Carbon Dioxide Emissions Factsheet*. Published at www.eea.eu.int/all-indicators-box (accessed 21 August 2002).

Food and Agriculture Organisation (2002). *PR 96/47 Fast Growing Cities present enormous challenges*. Published at www.foa.org/WAICENT/OIS/PRESS-NE/PRESSENG/H41F.HTM (accessed 18 August 2002).

Figure 3.9

Figure 3.9 *Results from a survey of farmers selling their own produce at a local farmers' market in Peckham, London. The distance farmers travel and their fuel consumption is compared to the volume of produce they sell.*

Hayami, Y. and Godo, Y. (1995). *Economics and Politics of Rice Policy in Japan. A perspective on the Uruguay round.* NBER Working Paper No. W5341.

Johnson, S. (1756). Further thoughts on Agriculture, Universal Visitor. In *Johnson's Works* (D. J. Green, ed.) vol X, Political Writings. Yale (1977).

Kol, R., Bieiot, W. and Wilting, H. C. (1993). *Energie-intensiteiten van voedingsmiddelen.* Energy and Environmental Sciences Department (IVEM), State University of Groningen, Netherlands.

Kramer, K. J., Moll, H. C., Nonhebel, S. and Wilting, H. C. (1999). *Greenhouse gas emissions related to Dutch food consumption.* Energy Policy, **27**, 203–206.

Lampkin, N. H. and Padel, S. (1994). *The Economics of Organic Farming.* CABI Publishing.

Leach, G. (1976). *Energy and food production.* Institute for Environment and Development.

Nicholson-Lord, D. (1987). *The Greening of Cities.* Routledge, London.

Shorrock, L. D. and Henderson, L. D. (1989). *Energy use in buildings and carbon dioxide emissions.* HMSO.

Stanley, D. (2002). *Sustainability in practice. Achieving the UK's climate change commitments and the efficiency of the food cycle.* e3 consulting.

United Nations Conference on Environment and Development (1992). *Earth Summit '92.* Regency Press Corporation.

Vale, B. and Vale, R. (2000). *The new autonomous house.* Thames and Hudson.

Wright, S. (1994). *The handbook of organic food processing and production.* Blackwell Science.

4

URBAN AGRICULTURE AND SUSTAINABLE URBAN DEVELOPMENT

Herbert Giradet

TAKING STOCK

At the start of the new millennium we live in a world of unprecedented human numbers. There are currently about 6.3 billion people, and this figure is expected to increase to some nine billion by 2050. About half of the world's population live in cities, a figure which is likely to grow to two thirds by 2030. Most cities are being built on farmland, a factor that will certainly reduce the world's food production capacity unless city people produce significant proportions of their own food. So some important questions need to be answered: can the global environment cope with this 'age of the city'? Will there be any untouched natural systems left? How can the world's growing numbers of city people be fed?

A very useful methodology in this context is to measure the 'ecological footprint' of cities, drawing on the work of Canadian ecologist William Rees and his Swiss colleague Mathis Wackernagel. As they see it, we need to quantify the land areas required to supply cities with essential resource services – the footprints of cities. These consists of three main components – the surface areas required:

1. to feed cities;
2. to supply them with forest products; and
3. to reabsorb their waste, and particularly their carbon dioxide output.

Using this method, I have tried to assess the ecological footprint of the city where I live, London, which has a population of just over seven million. I found that it extends to around 125 times its own surface area of 160 000 ha, a total of 20 million ha. This breaks down as follows: London requires about 1.2 hectares of farmland per person, a total of 8.4 million hectares, or around 40 times its surface area. The forest area needed to supply it timber and paper is about 768 000 ha. The area that would be required to sequester its annual output of about 60 million tonnes of CO_2 is by far the largest – about ten million ha, or half the total footprint.

A key issue here is how worldwide urbanisation and growth in affluence will increase human demands for land surfaces. It has been estimated that if developing countries copy our Western urban lifestyles – in terms of demands for food, forest products and energy – we will need three planets, rather than the one that we actually have. It is of crucial importance, therefore, for cities in developed countries to become much more efficient in the way they use resources, and that certainly includes their food supply. Urban agriculture can make a crucial contribution here.

But here we are primarily concerned with the land area required to feed large cities such as London, and the fact that food now tends to come from further and further afield.

London, of course, was the world's pioneering mega-city. It grew from just under a million people in 1800 to about 8.5 million in 1939, a city of an unprecedented size. At first London drew on a largely local food supply. But new transport technologies made it possible to bring its food in from further and further away. Steam ships supplied grain from Canada and the USA, lamb from New Zealand, wine from France and Italy, oranges from Spain and Brazil, bananas from the West Indies and South America. Today food is brought to us from just about anywhere – no longer just in freighters and trucks, but also in aeroplanes flying in from half way across the world.

ENERGY AND LAND-USE

The site of Heathrow airport used to be London's market garden. Its sandy soil is very suitable for vegetable growing. Today, even though it is largely concreted over, Heathrow is still London's major food supplier, but in a rather different way: food is flown in from across the globe. Such a global harvest offers us great culinary variety, but it requires the availability of vast quantities of fossil fuels. By the time food that has been air freighted for thousands of miles reaches a London dining room table, it will have used hundreds of times as much energy as the calories it actually contains. But it is not only air freighted food that is tremendously energy intensive. Frozen fish from the Atlantic contain some 100 times more energy than their calorific value, meat from UK farms 50 times more. These are astonishing figures. It seems unlikely that we can sustain such a high-energy food urban system for long.

But it is not only the energy input into our food system that should concern us, but also its impact on global land use. Increased demand for meat, in particular, has become a primary cause of deforestation in virgin forests in the Amazon, Thailand, Malaysia and Indonesia. Animal feeds such as soybeans and manioc produced in these places have been used extensively in Europe and Japan for many years. Now, economic and urban growth in large developing countries such as China is causing a rapid increase in meat consumption there, and so increasing demand for soybeans from rainforest and savannah regions elsewhere. In Brazil, this process started in Mato Grosso on the southern edge of the Amazon. Massive new road building programmes are now under way in the Amazon itself, which will result in the conversion of ever-larger areas of virgin forest into soybean fields to supply the growing demand for soybean from China's cities.

Over the last 50 years, agriculture in developed countries across the world has been transformed into an ever more capital intensive, machine dependent system. In the UK, only one and a half per cent of the population is still producing food. Rural landscapes here no longer exists in their own right but for the sole purpose of supplying urban demands.

NUTRIENT FLOWS

Another important issue to be addressed is that, worldwide, we are seeing uni-directional food and therefore nutrient flows, from the country to the city, never to be returned to the land. This unsustainable system was pioneered in Rome 2000 years ago with the construction of the 'cloaca maxima' through which much of the city's sewage was flushed into the Mediterranean. More recently, in the 1850s, London, faced with major outbreaks of typhoid and cholera due to sewage pollution of the Thames, decided to separate its people from their sewage output. After much deliberation, London's authorities took the decision not to recycle the sewage, but to dump it in the sea instead. From 1858 – the year of the 'great stink' – onwards a huge investment in a sewage disposal system was made. But because most of the sewage was now flushed away rather than used as fertiliser, it became necessary to keep the farmland feeding London productive by artificial means. The age of artificial fertilisers and chemical farming had begun.

A few years ago, flying from Rio de Janeiro to Sao Paulo I observed Rio's vast, brown sewage 'plume' oozing out into the sea. Similar images can be seen at coastal cities all over the world. This one-way traffic of nutrients – from farmland, via cities into the sea – is causing havoc to coastal waters across the planet. These plumes contain the nitrogen, potash and phosphate that should be used for growing the crops we eat. In addition, a substantial proportion of the artificial fertilisers now used on the world's farms also ends up polluting rivers and coastal waters. If we are to create sustainable cities, we need to look at the nutrient flows between cities and the countryside, and at the 'metabolism' of our cities.

RELEARNING URBAN AGRICULTURE

Urbanisation, and the shift from rural to urban living by billions of people, has not only resulted in major environmental problems, but also in urban poverty, food insecurity, and malnutrition, particularly in developing countries. But, almost unnoticed, it has resulted in the growth of a remarkable phenomenon: urban agriculture. According to UNDP, in 1996 some 800 million people were engaged in urban agriculture worldwide, with the majority in and around Asian cities. Of these, 200 million were thought to be market producers, with 150 million people employed full-time. Urban agriculture now occupies large minorities in many cities around the world. Cities such as Havana, Accra, Dar-es-Salaam and Shanghai have been studied extensively. But in thousands of other cities around the world people are also quietly getting on with their own food production.

Over the last several years I had the opportunity to witness urban agriculture in many different parts of the world. I was interested in this because, among other things, urban agriculture can help cities make the best possible use of organic waste materials.

Urban agriculture is conducted within or on the fringe of cities. It is concerned with growing plants and herbs and the raising of animals for food and other uses. The production of tree seedlings, ornamental plants and flowers is also part of the picture. It is becoming apparent that in order to survive in a globalising food system, urban farmers must be highly innovative and adaptable. They must be able to cope with city constraints and tap as effectively as possible onto urban assets and resource flows. For instance, an important component is the use of compost and manures available in the urban environment.

Despite globalising tendencies, local urban food production is being practised in many places. In recent years its importance has been increasingly acknowledged by researchers, politicians and urban planners – transforming it from a largely neglected activity, to a major force for creating sustainable livelihoods for urban people.

In the developing world in particular, urban agriculture can greatly contribute to urban food security, improved nutrition, poverty alleviation and local economic development. In developed countries it is recognised as contributing to the reduction of 'food miles', involving city people in food growing and distribution via farmers' markets.

Urban agriculture often builds on ancient traditions. Historically, most cities grew out of their own hinterland, and some contemporary cities are still deeply 'embedded' in their local landscapes, even in Europe. For instance, Florence is still surrounded by orange and olive groves, vineyards and wheat fields on which a large proportion of its food requirements are grown. Many cities in Italy, and also in France, still have very strong relationships to their immediate hinterland, with 'peri-urban' agriculture still much in evidence.

I found the same in China. China has an age-old tradition of settlements permeated with food growing areas. Today, at a time of very rapid urban-industrial growth, urban agriculture is still a very important issue for the Chinese. Even mega-cities such as Shanghai, one of the fastest growing cities on the planet with about 15 per cent growth per year, maintains its urban farming as an important part of its economic system. A major shift has taken place, however, from 'intra-urban' to 'peri-urban' agriculture. As housing and office developments grew within the city, farmland there has been lost and food growing has shifted increasingly to the city's periphery.

The Shanghai city authorities administer an area of about 600 000 hectares of land: 300 000 hectares of this are built up areas within the city itself. But as a deliberate policy, some 300 hectares of land on the edge of Shanghai are now deliberately maintained as farmland for feeding the city. Most of this land is used to supply rice and wheat, though, as we saw above, production of animal feeds such as soybeans, particularly for beef produced in US-style feedlots, now takes place increasingly in far-flung places such as the southern Amazon.

Tens of thousands of hectares on the outskirts of Shanghai are intensely cultivated with a great variety of vegetables. The Chinese like to cook fresh, locally grown vegetables. Stir-frying wilted vegetables is not regarded favourably. Glass and polythene greenhouses are now much in evidence, producing three to four successive crops a year in Shanghai's warm climate.

On the outskirts of Beijing, too, vegetable cultivation is much in evidence. But farmers have had to develop ingenious systems to cope with the much colder climate. Greenhouses, too, are much in evidence. During frosty conditions in January and February, they cover their polythene tunnels with several layers of bamboo mats in the evening to keep the heat in at night. Few growers in and around Beijing use coal fired heating systems in their greenhouses to cope with the icy conditions outside.

In Chinese cities 'closed-loop' systems, using night soil as fertilisers for urban vegetable growing, are still widely maintained. The night soil is diluted, perhaps ten to one, and then ladled onto vegetables beds. I was told that people prefer vegetables grown with night soil fertiliser because they taste better. But most new apartment and office buildings, which are in evidence everywhere, have water closets, and it remains to be seen whether appropriate ways of using wastewater in urban farming can be developed.

In Russia, too, peri-urban food growing is an age-old tradition, with many people retreating to their dachas at weekends to cultivate crops in highly productive gardens. In St Petersburg most people are involved in urban farming: there are some 560 000 plots being cultivated on the periphery of the city. Even in remote places such as Irkutsk in Siberia with its very short growing season, I have seen people cultivate an amazing variety of vegetables, including cucumbers and tomatoes, in well-insulated greenhouses, both for home supply as well as for sale in markets.

In South Africa, of course, during the apartheid days it was forbidden for the black majority to farm land within and around cities, because that meant people were there to stay. But now a dramatic growth of urban agriculture is under way as people get a permanent foothold in their towns and cities. And throughout Africa, in Ghana, Kenya, Tanzania and elsewhere, much food growing takes places within cities, because they are often still very low density and there is room for food growing. Women tend to be the cultivators in urban areas.

Havana in Cuba is a particularly remarkable example of urban agriculture development. As a result of the collapse of the Soviet Union, Cuba lost a large proportion of its sugar export earnings. So a few years ago the authorities decided to practise food import substitution and to encourage urban agriculture within the city. Composted bagasse from the sugar cane fields is often used as fertiliser. Sugar cane, ironically, is grown with artificial fertilisers, but the bagasse that is composted effectively becomes an organic fertiliser and this is used on raised vegetable beds called 'organiponicos', which are irrigated with pumped underground water. In Havana, some 20 000 people now grow fruit and vegetables, mainly on plots adjoining their apartment blocks.

Whilst urban farming is being recognised more and more as an important source of food and income generation in cities around the world, adequate institutional frameworks at national, municipal and local levels are still often lacking. It is becoming important to find ways to overcome this obstacle. Rather than competing with rural agriculture, urban and rural agriculture should be seen as complementing each other since urban agriculture tends to focus on products that require closeness to the urban markets such as vegetables, flowers, poultry and eggs.

Opposition to urban agriculture has tended to come mainly from public health and urban planning circles because of concern about water pollution and soils contaminated by heavy metals. However, research has shown that concerns about adverse effects on public health have been exaggerated. There is broad consensus now that urban agriculture is an important area for government support at national as well as municipal level.

Developed countries

But, anybody who thinks that urban farming is only a phenomenon primarily of poorer countries, should have a look around parts of New York City. In the Bronx, for instance, an astonishing range of vegetable gardens sprang up in the 1980s, primarily in areas where drug-related gang warfare resulted in houses being burned down and gardens left abandoned. With the help of people from the New York Botanical Gardens, local people turned dozens of vacant lots into thriving vegetable gardens. Many also grew crops for the sake of their children who they wanted to teach about growing vegetables and keeping chicken and rabbits.

In California, too, urban farming is widely practised. In the university town of Davis, some enlightened developers some years ago decided to build a 'permaculture' suburb. They surrounded new ecohouses with vegetable plots and orchards. Even good quality wine is now produced right in the middle of Davis.

In the USA, the growth of farmers' markets has been a remarkable phenomenon in recent years. Despite the enormous dominance of supermarkets, farmers' markets have been an extraordinary success, not only in California, where the growing conditions are best, but also in New York. There are now over 3000 farmers' markets. In the UK, too, there has also been a resurgence of farmers' markets, from nothing about ten years ago to about 300 in 2002. And allotment growing has maintained its popularity within cities, though today it is less and less done by retired men, but increasingly by women who want to grow some of the vegetables for their families.

In the UK there is also urban food production. For instance, in Nazeing in Essex, just outside London, one can see how farming has come under pressure. Like Heathrow, Nazeing used to be a major centre for vegetable growing, in a landscape full of greenhouses. But few growers could compete with cheap, imported vegetables and many had to abandon their plots. The few that are left now grow only one crop: cucumbers. These are grown hydroponically in greenhouses that look as clean as operating theatres. The growers are mainly second generation Italians. That is because the people who used to own these greenhouses couldn't make them pay any more. The Italian prisoners of war who had been their labourers during and after the war, took over the last remaining greenhouses, partly because they could draw on additional supplies from Italy. When it isn't cost-effective to grow cucumbers in the winter in England, they truck them in from Italy instead.

On the outskirts of Bristol, attempts have recently been made to set up new organic market garden schemes. For instance, at Leigh Court outside the city, an organic vegetable box scheme was set up in the 1990s. But it is difficult to compete with cheap, imported crops – three quarters of the organic vegetables consumed in Britain are actually trucked and flown in from elsewhere, at great energy cost. But some initial steps to revive peri-urban agriculture are now being taken.

CITIES AS SUSTAINABLE SYSTEMS

Urban agriculture is an important aspect of the wider issue of urban sustainability, both by being able to supply food from close-by and by offering livelihoods for city people. Another important issue, as already discussed, is the efficient use of nutrients from the urban metabolism that would otherwise end up as pollutants in rivers and coastal waters.

In many cities attempts are being made to use wastewater in urban food production. This applies particularly to cities in hot and dry places. For instance, in Adelaide, Australia, tens of thousands of hectares of land on the edge of the city are cultivated using wastewater from the city for irrigation, growing vegetables as well as grapes and fruit. There is some concern about trace quantities of heavy metals that could accumulate in the soil, but it would take decades to cause any problems. Adelaide's wastewater crop irrigation system is regarded as one of the great success stories of urban agriculture.

In Bristol, Wessex Water has developed its own system for turning sewage into a soil conditioner and fertiliser. It dries the city's entire sewage output and turns it into small pellets called Biogran, which are then sold to farmers and land reclamation companies. Again, trace amounts of heavy metals have been quoted as problematic. But this is becoming less of a problem because cars no longer run on leaded fuels in the UK, and in Bristol de-industrialisation has led to a great improvement in the quality of Bristol's sewage sludge.

Another important aspect of sustainable urban development is the creation of new kinds of eco-efficient housing estates. This concept is now flourishing across Europe. In South London, a pioneering project was completed in 2002 – the Beddington Zero Energy Development. This is a housing and workshop project for some 200 people, created by the Peabody Trust and the Bioregional Development Group. All buildings have south-facing facades and 30 centimetres of insulation in walls, floor and ceilings. The apartments require only 10 per cent of conventional heating energy and this is provided by a small, wood-chip fired combined heat and power plant. There are solar electric panels installed on all the south-facing facades and these will supply electricity to a small fleet of electric cars. All apartments have their own small roof gardens that can be used for recreation and/or vegetable growing. The estate's wastewater is treated in a 'living machine' which uses plants and zooplankton for extracting surplus nutrients. The water from the treatment plant is used for irrigating gardens.

The Beddington project shows how ideas for making cities eco-efficient can be turned into practical reality. We need to turn the linear throughput of resources through our city into circular systems, where minimal inputs into the city result also in minimal waste outputs. Energy efficiency, resource productivity – these are the key themes in this context.

Good urban design in the twenty-first century should start by mimicking natural eco-systems. Above all else, we would be well served to learn from the metabolism of natural, closed-loop systems in which all wastes are recycled into resources for future growth. This is an issue for policymakers, but also for the general public which needs to exert pressure on local and central governments, and on developers, to adopt forward looking practises.

We are certainly seeing a lot of interest in these ideas now in cities all over the world. But a major push is still needed to reduce the wasteful resource consumption and the vast, sprawling ecological footprints of cities that we have at present. We need to move towards much more localised, efficient, circular urban systems, and this scenario certainly includes the use of land within and on the edge of cities for food production. Ideas for creating sustainable cities have been around for some time, but implementation on an adequate scale has hardly begun.

5

FOOD MILES

Angela Paxton

INTRODUCTION

Consumers in rich industrialised countries are accustomed to being able to choose from the global breadbasket whenever they go to the local supermarket. A UK shopper can buy grapes from Chile, green beans from Kenya and bottled water from Canada. However these food choices do not come cheaply: there is a whole range of environmental, social and economic costs which result from the increasing long distance trade in foods.

The distance food travels, as well as being environmentally damaging directly through transport emissions, adversely affects the way food is grown and treated on its journey along the various links in the food chain. There are also impacts for food producers, particularly small farmers and growers trying to compete in a global marketplace, and for consumers, who are eating more long distance, treated food and products rather than fresher, local foods.

The separation of consumption from production results in the psychological distancing of food producers from consumers, with shoppers having little idea about the many implications of their food buying habits. This consumer ignorance allows yet more abuses of the environment, people and animals, which might not be tolerated if they were happening closer to home.

This chapter will highlight the impacts of our transport-dependent food system, look at the forces behind the food miles and explore what can be done to reduce and alleviate the impacts of long distance food distribution.

THE FOOD MILES FOOD CHAIN

Agriculture

Long distance trade in foods leads to specialisation in farming and the use of local resources for export to other regions or countries rather than for local consumption. Thus crops will be grown over large areas in industrialised conditions to achieve economies of scale and cut costs in order to compete with growers elsewhere. Producers will concentrate on growing a small number of crops, using varieties specified by food manufacturers or retailers for particular qualities, such as their ability to withstand long periods in transit and storage, or their suitability for processing. Crops grown in monocultures are easy prey to pests and diseases, which quickly become resistant to agro-chemicals, and farmers find themselves on a 'chemical treadmill', having to apply increasing amounts of pesticides and herbicides.

Transport

Essentially, food miles can take two forms: transport within a country as a result of the centralised distribution systems of large retailers and food manufacturers, and transport between countries as nations import and export fresh and processed food products. Within the UK, food, drink and tobacco were responsible for a third of the growth in road freight between 1978 and 1993. Meanwhile the transport of food and drink products around the UK increased by 50 per cent in the past 15 years, while the amount of food being transported increased by only 16 per cent, see Figure 5.1.

Figure 5.1 *Comparison of the average distance and total quantity of food transported in the UK between 1978 and 1999.*

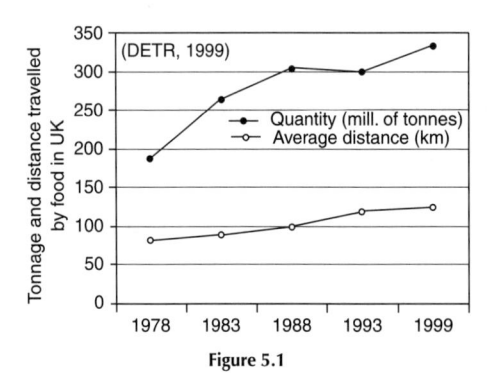

Figure 5.1

Internationally too, food transport is on the increase. UK farmers rely on exports while British consumers buy increasing amounts of imported food. The UK trade gap in food and drink increased from £3.5 billion in 1980 to £8.3 billion in 1999. The UK now imports a variety of foods, such as apples, carrots and onions that can be grown here, even when they are in season. Countries often import and export substantial quantities of the same food products: in 1997 the UK imported 126 million litres of milk and exported 270 million litres, see Figure 5.2.

Air freight

Air freight of fresh foods has doubled in the past twenty years, a trend which is set to continue. Air transport is particularly damaging to the environment as it results in 37 times more carbon dioxide emissions than sea freight, and because pollutants are emitted at high altitudes where more damage is done in terms of ozone depletion and global warming.

Driven to shop

Shopper miles are also an issue as more people are driving to out- and edge-of-town supermarkets to do their food buying. One shopping trip by car can use more fuel than freight transport up to the point of sale, even if the produce is imported (athough not by plane).

The 3 'P's: processing, packaging, pesticides

Food is perishable and needs to be preserved for long distance transport, to prevent spoilage and contamination. Common methods of preserving food include processing, packaging and pesticide spraying.

Processing

The manufacture of processed foods from raw ingredients is an energy intensive procedure using up to ten times the energy needed to grow the crop in the first place. Processed foods are likely to have incurred greater food miles than fresh produce because ingredients and packaging materials will be sourced from other parts of the country and abroad. Food manufacturers are increasingly centralising production, some having just one factory to serve the whole EU, in order to cut costs.

A study by the Wuppertal Institute found that to bring one truckload of 150 g strawberry yoghurts to south Germany, strawberries came from Poland, yoghurt from north Germany, corn and wheatflour from the Netherlands, jam from west Germany and sugar beet from the east of the country, while the aluminium cover for the strawberry jar came from

Figure 5.2 *Comparison of food imports into the UK between 1961 and 1998.*

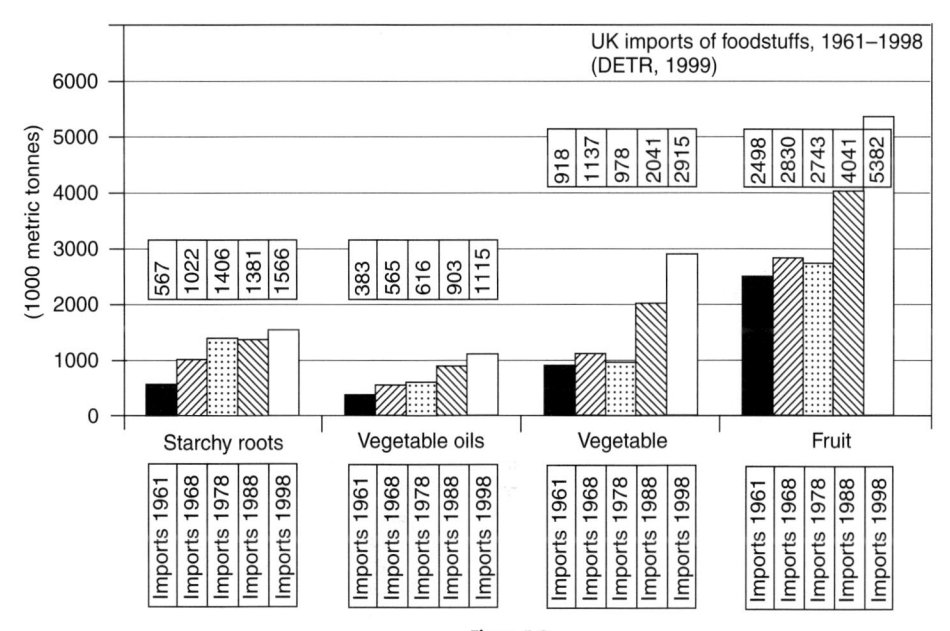

UK imports of foodstuffs, 1961–1998 (DETR, 1999)

Category	1961	1968	1978	1988	1998
Starchy roots	567	1022	1406	1381	1566
Vegetable oils	383	565	616	903	1115
Vegetable	918	1137	978	2041	2915
Fruit	2498	2830	2743	4041	5382

(1000 metric tonnes)

Figure 5.2

300 km away. Only the milk and glass jar were produced locally.

The basic components of many processed foods, such as cakes or biscuits, are the same, so the apparent choice and variety in supermarkets is exaggerated. Consumers are persuaded to pay more via 'added value' for cheap ingredients which are made to seem appealing through the use of additives, glamorous packaging and aggressive marketing.

Pesticides

As well as the increasing need to apply agrochemicals when crops are grown in industrial scale monocultures, pesticides are applied to fresh produce to preserve it during long distance transit. For example, 85 per cent of Cox apples in the UK are treated with post-harvest chemical dip or drench prior to storage. Unlike pesticides applied in the field, these chemicals are designed to stay on the produce. The Department of Health recommends peeling orchard fruit to avoid eating pesticide residues.

Packaging

Packaging enables foods to withstand long periods in transit and glamorises the contents of processed food and drink products. Food distribution requires at least four types of packaging, including primary packaging on the food product, secondary packaging, such as cases or boxes, transit packaging such as crates, and carrier bags or boxes to carry foodstuffs from shops to the home.

1.5 billion dustbins of packaging waste are produced in the UK each year, most of this ending up in landfill sites. Up to a third of this waste is used to protect food and drink products. Long distances between producers and consumers make reducing and reusing packaging much more difficult, whereas local food systems can reduce the need for packaging and makes the reuse of containers much more viable. Returnable bottles, best suited for local distribution systems, typically use one quarter of the energy of any single trip package.

IMPLICATIONS OF THE FOOD MILES CHAIN

Environment

Air pollutants and climate change gases are released as fossil fuels are used for production, transport and packaging of food and drink products. The CO_2 emissions created by producing, processing, packaging and distributing the food consumed by a family of four come to about eight tonnes a year. Road freight alone is responsible for a high proportion of toxic emissions, such as nitrogen oxides and volatile organic compounds, which are implicated in a number of public health problems, such as asthma and other respiratory diseases. Meanwhile, industrial agriculture supplying international markets involves overuse of pesticides which leach into groundwater and threaten wildlife, land erosion, loss of wildlife sites and reduced biodiversity in wild and cultivated species.

Biodiversity

Growing food for sale in distant markets requires a high level of specialisation to compete with producers in other countries around the globe. Only those varieties will be grown which can withstand long distance transport and storage and look good on the supermarket shelf, rather than the local, tasty or nutritious varieties. For instance, 75 per cent of strawberries produced in the UK are a single variety, Elsanta, which doesn't taste particularly good, but has a long shelf life and good travelling characteristics. However, it is susceptible to diseases such as powdery mildew, which necessitates the use of chemicals such as methyl bromide, a toxic chemical which destroys the ozone layer. Official testing found pesticide residues on 88 per cent of both imported and UK strawberries.

Pressure for specialisation and modern farming methods mean that many crop varieties are disappearing: there were 287 varieties of carrot in 1903, but now only 21. This poses a threat to food security as the genetic base of our crops is now narrower than it has ever been, increasing the likelihood of widespread crop failure.

Small producers

Industrial, specialised agriculture favours the large over the small. Small producers in the UK and developing countries are squeezed out of the marketplace as they are unable to produce crops cheap enough and in sufficient quantity for supermarkets, food processors, and transnational corporations, who prefer to deal in large quantities. These companies add considerable mark-ups to foods, while continuing to pay farmers low prices. One UK farmer traced his three year bullock to find out the profit margins when he lost £32 on a sale in 1998. He found that between the abattoir and final sale a profit of £800 was made on the animal.

Developing countries

Up to 1970, food security was seen as an important objective in food production in developing countries. More recently, food and agricultural products have come to be seen as commodities for trading, to pay off debts and earn foreign exchange. However, producing for distant markets promotes insecurity for growers, and the lowering of environmental and social standards to cut costs and compete in international markets. For example, many pesticides are used in developing countries on crops destined for industrialised countries, which are banned in the latter because of environmental or health effects. According to the United Nations Environment Programme, at least 40 000 people are killed each year and up to one million people are made ill or permanently damaged by misuse of pesticides, mostly in developing countries. Even in the UK, thousands of farmers and farmworkers believe they have been poisoned by organophosphorous sheep dip, a close pesticide cousin of nerve gas.

Public health

Over-processed, over-preserved and over-packaged foods mean that consumers are buying foods of low nutritional value. Diet related diseases such as heart disease, diabetes and appendicitis increase as societies move towards the Western diet, low in fruit and vegetables, and high in refined starches, fats and sugars. Pesticide residues pose additional health risks, although it is the farmworkers who are most likely to suffer immediate health effects rather than the final consumers.

Local economies

When consumers buy food from supermarkets, almost all the money is lost to the local economy. It may be lost to the national economy if the food is imported, or the product of food companies based abroad. Only a small amount of the money spent will stay in the area, in the form of wages to local people working in the retail outlet. However, every £10 spent on local food through small initiatives is worth £25 to the local food economy, because of the 'multiplier' effect, where money is spent on other local goods and services.

FORCES BEHIND THE FOOD MILES

Cheap fuel and transport

The forces behind food miles are complex. A key area is the low costs of transport and fuel which do not reflect the full environmental and social costs of their use. Transport is responsible for not only local and global air pollution, but for contributing to health problems, climate change and destruction of the ozone layer. It also creates noise pollution, vibration, fumes and dirt, accidents, wear and tear on transport infrastructure and the destruction of wildlife habitats. These costs are borne by society and the environment.

Middlemen

The 'middlemen' in the food chain, such as freight operators, processors, packaging companies and retailers, benefit most from increasing food trade and transport. On a national level, the major retailers in the UK control over 80 per cent of food sold in the UK. Supermarkets have shelf space to fill all year round and import foods to fill seasonal gaps, even when those foods are in season in this country. The centralised distribution systems and 'just-in-time' ordering systems of supermarkets result in increased freight transport, as journeys are made when goods are ordered, not when lorries are full. This results in more journeys being made than necessary to transport a given amount of goods.

Internationally, transnational corporations (TNCs) exploit land, labour and resources in developing countries for the production of cheap export crops. Considerable mark-ups are added to these products before being sold to rich consumers in the north, the profits being reaped by the TNCs. Increasingly, added value food and drink products, such as soft drinks, are sold back to consumers in developing countries.

International policies

In order to earn export earnings, IMF and World Bank policies have encouraged, through structural adjustment programmes (SAPs), more and more farmers in developing countries to move away from the production of food for local consumption to producing similar 'cash crops' such as coffee, tea, and horticultural products for export. The result has been over-supply of these products, causing the value of food commodities to plummet, bankrupting many producers. There is still intense pressure to further liberalise trade in foodstuffs. In World Trade Organisation (WTO) negotiations, the Cairns Group is seeking to remove barriers to international trade in food and agricultural subsidies.

Consumer ignorance

Food corporations argue that consumers drive the food industry. Indeed, consumers have become used to being able to buy all foods at all times of the year, regardless of seasonality, and at cheap prices. Consumers are buying food produced by people who they do not know, and probably will never meet, making them less concerned about their welfare. This lack of consumer loyalty adds to the fickleness of westerners' food choices, rendering them easy prey to the marketing strategies of multinational corporations, and making farmers' livelihoods increasingly insecure.

The geographical separation of consumers from food production renders shoppers ignorant of the many abuses of the environment, farm workers and farm animals, which might not be tolerated if they were going on in a neighbouring field. However, demand for fair trade and organic foods is now a significant force in the UK food market – organic sales in this country are increasing by 40 per cent per year – reflecting that many consumers do care about how their food is produced.

SOLUTIONS

To address the multiple issues around food miles, change needs to happen on many different levels, from new consumption habits to government policy. On an individual level, consumers can choose to buy local, seasonal or fairly traded foods and ask shops and supermarkets to stock these products. Another option is to grow food in gardens or allotments.

Groups of individuals can set up local food schemes, such as community growing initiatives, vegetable box schemes and farmers' markets. Buying direct also re-establishes the links between producer and consumer and can mark the beginning of a constructive dialogue towards more diverse and organic production. In particular, consumers can ask their local farmer to use more sustainable production methods and to grow a greater range of crops to satisfy local demand for diversity. Local authorities can support such initiatives through funding, advice and making land available for community food schemes.

National government can introduce policies to internalise the environmental and social costs of transport, such as introducing a weight distance tax for air and road freight. Compulsory food miles labelling for products should be introduced and advice and financial assistance given to direct marketing schemes. Food From Britain, the dti and industry sponsored body which promotes exports of foods, should be transformed so that its emphasis is on import substitution. Aid and debt relief for developing countries should be linked to sustainable development initiatives such as diversification and sustainable agricultural production. Pressure needs to be exerted for the multilateral adoption of minimum standards for working conditions, environmental protection and animal welfare in the production of goods and services, at regional and international levels.

REFERENCES

DETR (1999). Energy and Environment Transport Statistics. HMSO.

SAFE Alliance (now Sustain) (1994). The Food Miles Report: the dangers of long distance food transport.

Sustain (2001). Food miles: still on the road to ruin.

Sustain (2001). Eating Oil: Food supply in a changing climate.

Plugging the Leaks project. New Economics Foundation, London. (accessed November 2004 http://www.pluggingtheleaks.org/)

6

SANDWELL:
A RICH COUNTRY AND FOOD FOR THE POOR

André Viljoen

Architects and urbanists are used to viewing the city through plans. Periodically, new means of viewing the city become available which help us to understand the spatial consequences of social phenomena. In the early part of the twentieth century aerial photography became available and made clear the effects of rapid industrialisation on the urban fabric.

Today we have a new way of reading the city with plans. No longer figure and ground plans, but plans generated using geographical information systems (GIS). GIS allows data to be mapped spatially, presenting layers of information, which by correlation can map, for example, access to resources. In terms of sustainability, this begins to let us judge the equality, or otherwise, of resource availability.

An example of the use of GIS mapping to record access to healthy food is provided by a study entitled, 'Measuring Access to Healthy Food in Sandwell', undertaken in the year 2000 (The University of Warwick and Sandwell Health Action Zone, 2001). Sandwell is located in the Midlands region of England.

Figure 6.1 presents results from the Sandwell study; these maps graphically illustrate the lack of access to affordable healthy food in an area suffering from poverty. Mappings like these provide new readings of the city. They deal with a reality of occupation that goes beyond the physicality of traditional maps. They present the designer and citizen with an objective view of contemporary inequalities.

The Sandwell study makes startling reading. The project had three aims, to produce indices of access to food in the deprived area of Sandwell, to examine how such maps could help in the development of strategies to promote healthy eating among low income households and to work with local retailers to improve access to healthy food.

The area in which the study took place had previously been identified as one of high socio-economic deprivation and poor health outcomes, and had been targeted for Department of Health promotional initiatives. The Sandwell study found that large networks of streets and estates had no shops selling fresh fruit and vegetables. In areas where fresh fruit and vegetables were available it was often of poor quality and expensive. As the maps in Figure 6.1 show, inexpensive good quality fresh fruit and vegetables are sold in small concentrated areas, but these areas require the use of public or private transport for the majority of the population to reach them. Competition from supermarkets has reduced the supply of local fresh fruit and vegetables.

The implications of this study, as recorded by its authors, are quoted in full, as they provide a powerful argument in favour of a shift in policy towards the introduction of productive urban landscapes supporting local food production.

Implications from the study, 'Measuring Access to Healthy Food in Sandwell':
1. Poor health, deprivation and unhealthy eating patterns in Sandwell are strongly interlinked.
2. Eating patterns in Sandwell may be determined by socio-economic and geographical factors rather than real choice or knowledge.
3. Tackling food access through the use of volunteer labour is not the solution.
4. Good public transport can reduce but not remove the problem of food access.
5. There are economic, social and environmental reasons to develop a highly localised food economy that is more sensitive to the needs of Sandwell's people.

Figure 6.1 requires further examination. The maps show roads that are within 500 m of a postcode containing shops selling food. 500 m is considered the distance a fit person can walk in ten to fifteen minutes, someone with children and shopping bags would take longer. This distance has been used in a similar study located in London (Donkin et al., 1999) and is considered a reasonable maximum distance for people to walk to shops. Postcodes are used to locate shops accurately using a GIS mapping system. In England a post code usually contains about 12–14 dwellings or addresses, which correspond to a small area.

The part of Sandwell studied has many households on low incomes, with limited shopping and transport networks. The local shops which do exist in the area, cannot sell fresh healthy food as cheaply as large retailers or markets, as they have low levels of sales because their customers are poor. Large retailers do not locate in these areas because of the poor

SANDWELL MAPS
measuring access to healthy food

Road Data Ordnance Survey
© Crown Copyright all rights reserved

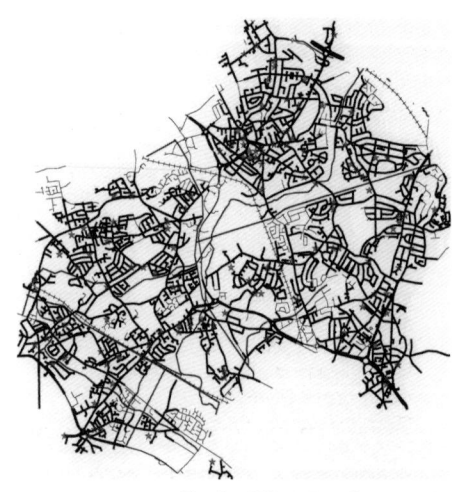

LEGEND

Roads within 500 m

Roads further than 500 m

Railways

Canals and streams

★ Shops

Roads within 500 m of a postcode containing one or more shop selling some food.

Roads within 500 m of a postcode containing one or more shops where food is reasonably priced and which sell at least eight kinds of fresh fruit and vegetables.

Figure 6.1

transport facilities and a poor population. The Sandwell study found that food, which is readily available in the area, tends to be less healthy; 'high fat, high salt, cheap easily storable foods.' (ibid, p.11). Thus a vicious circle is created, which results in limited access to affordable fresh fruit and vegetables.

When comparing the maps in Figure 6.1, the extent of pedestrian exclusion to areas selling affordable fresh fruit and vegetables is obvious. It could be argued that this exclusion is only relevant within areas of poverty, as in wealthier locations the population has access to better transport facilities, in particular private vehicles. This argument loses its validity if one accepts that a goal of sustainable development is to reduce reliance on unnecessary transport. Thus the lessons from the Sandwell study are generally applicable if one wishes to progress towards environments which support equitable, sustainable development.

The introduction of productive landscapes to an area like Sandwell would answer many of the problems observed. Food would be produced locally, seasonally, it would be fresh and the creation of market gardens within Sandwell would provide employment. By creating corridors of continuous landscape, within which the urban agriculture fields are located, transport by foot and bike could be improved, as well as providing residents with access to urban nature. These are radical proposals that go to the very heart of governance. The authors of the Sandwell report recognise this when they state that 'food access has to be part of the mainstream national and regional level policy agenda for area regeneration, and for tackling poverty and social exclusion and reducing inequalities in health' (ibid, p. 8).

If this inclusive approach is to work it will be necessary to reconsider the role of local government as an agent actively planning for productive urban landscapes. Such notions go against the current hands-off policy of local government, which appears to view any specific planning objectives, if associated with an approach that limits the operation of the 'free market' as fundamentally flawed. As the Sandwell report indicates, the market has not been able to deal with poverty. This is a good example of where top-down management is required to facilitate a locally driven process of regeneration, which responds to local conditions and has an agenda of improvement rather than aiming to manage the status quo. The role and importance of local government in promoting sustainable development must be strengthened. If fundamental issues such as sustainable food production are to be explored and achieved, then it is clear that the 'market' left to its own devices cannot address inequality.

The spatial design and social implications of introducing local food production within productive landscapes are enormous. Designers must realise that the necessity and justification for new visions lie beyond our own disciplines, and within the new territories being identified by a critical observation of failings within the status quo.

REFERENCES

The University of Warwick and Sandwell Health Action Zone (2001). *Measuring Access to Healthy Food in Sandwell*. Sandwell Health Action Zone.

Donkin, A. J. M., Dowler, E., Stevenson, S. and Turner, S. (1999). Mapping access to food at a local level. *British Food Journal*, **101**(7), 554–564.

7

PLAN IT:
AN INCLUSIVE APPROACH TO ENVIRONMENTALLY SUSTAINABLE PLANNING

Dr Susannah Hagan

Increasingly, the environmental case for urban compaction is taken as given. It is even rolled out to defend the perfectly straightforward commercial exploitation of valuable inner city sites, for example, at London Bridge. Suddenly this intensification is beneficial, not only financially, but environmentally, and the injection of thousands more people through a hypodermic high rise into an already overstretched infrastructure is left unexamined – at least by those standing to benefit from such development.

Others, however, are questioning compaction as necessarily beneficial environmentally. The concepts of 'productive landscape' and, more specifically, 'urban agriculture' are indicative of such questioning, and can be part of a very different way of conceiving of city and non-city. It requires thinking about the unbuilt as potentially an event of equal intensity to the built, where the built is indicative of cultural intensity, and the unbuilt of ecological intensity. The unbuilt, the uncompacted, can be viewed, not as a 'waste of space', but as productive space, space that is 'used' in different but equally valuable ways from building on it.

Since the Second World War, northern Europe and the USA have consistently lost urban population to the suburbs and beyond. Though this trend is now stabilising, and in some cases reversing, it is still difficult to keep families in cities in these parts of the world. It may one day be equally difficult in the developing world, as its urban populations mature and become more prosperous, able to choose where they want to live. The drift away from cities in areas of the West is only in part to do with cost, crime and inadequate schools. It also has to do with space, particularly space for children. In pushing for higher densities in existing cities, two things happen: more of the very thing people seek is lost – space – and attention and resources are concentrated on the city at the expense of the suburbs and the countryside, which are also problematic environmentally.

Urban compaction requires higher densities, but higher than what, and when is higher too high? This is impossible to quantify. What is desirable for one culture is intolerable for another, and what is chosen by one economic class might be another's lack of choice, because the desirability of increased densities relies heavily on how they are designed and maintained. Few Londoners would find the density of Kowloon acceptable, but then given a choice, a resident of Kowloon may very well not find it so either. In part this is because as densities increase, so too does the need for decompression space. Which returns us to the idea of unbuilt space and its relation to urban compaction, since it appears, even after a cursory discussion like this one, that compact-is-good is only good if it can at the same time incorporate space as well – space as parks, sports fields, squares, gardens, allotments or whatever.

Environmentally active urban space needs trees to clean and oxygenate the air. This in itself is productive – of air quality and the lowering of fossil energy demand to counter adverse environmental conditions. A virtuous circle. Does environmentally active urban space also require urban agriculture? Like most things environmental, it depends – I'd suggest more on the culture than the economic system. A city full of people who have recently migrated from farms, as in China, will find it easier to cultivate urban land than a population that has been urbanised for generations and doesn't know one end of a carrot from the other. So, too, a population accustomed to communal ownership, and/or communal activity, might find the idea easier, as urban land would often have to be shared and developed by groups of cultivators.

The 'who' and the 'where' of urban agriculture are therefore crucial. I don't think anyone is seriously suggesting that individuals start planting cabbages in Manhattan. I hope not, as trees would be much more environmentally 'productive' in such a context. But once you begin looking elsewhere than at very dense nodes with very high land values, urban agriculture begins to look more viable. It could also begin to include peripheral and suburban agriculture as well, and vitally so. As a term, 'urban agriculture' is a vivid description of what it aims to do, but it is also a limiting one. For an ostensibly radical idea, it continues the western tendency to create meaning and make decisions through the establishing of binary opposites. 'Urban agriculture' is innovative as it stands in contrast to 'rural agriculture'. Consequently, the interstices are left out of this opposition – the often slack and inchoate outer rings of cities and the suburbs beyond, with underused parks that produce a sense of unease which guarantees further underuse, and rubbish tips, and low rent industrial parks and warehousing, and derelict lots. In the case of London, there is also the green belt, a vast divide between city and countryside, but an empty divide, like a ditch or a moat, an absence rather than a presence.

If we were able to think of this and the other spaces I've described being potentially as intense, if not more intense, than the areas of settlement with which they are interlaced, then urban agriculture becomes one of several strategies for intensifying, without necessarily compacting, one among many landscape interventions. For instance, more woodland for biofuel, dumps turned into all-weather ski slopes, reed beds for filtration, fishing and nature centres, plant nurseries, weekend camping sites, walking trails, and model farms, so that people from the city and the suburbs and the countryside are as likely to mix and participate here as they are at an event in the city centre or some country town. This productive land, some of it more productive environmentally than socially, some the reverse, would be woven in and out of the city, the suburbs and the countryside, with urban agriculture an important thread in this; important because in this country, at least, the demand for organic produce outstrips its local supply. Urban agriculture, as I've extended its catchment area, would be well placed to contribute to this local supply – organic market gardens that would be part of a larger system of circular production and consumption – and it would be well placed to knit up types of settlement typically viewed as mutually contradictory, through its appearance in all of them.

The acquisition of appropriate sites, and the cultivation of those sites, would require considerable organisation, perhaps on the basis of some kind of charitable trust, or as a local or regional government initiative. But then, so would any intervention seeking to turn unused or underused land towards environmentally productive uses, like biofuel or wildlife habitat. The essential prerequisite of any attempt at change in this direction is that we begin to conceive of city, periphery, suburb and countryside, not as discrete, and for the most part, hostile categories, but as parts of a continuum that stretches from the most densely inhabited areas that are least active ecologically, to the least densely inhabited areas that are most active ecologically. I mean by 'ecologically' inclusive of all conditions and scales. For urban agriculture to work, it is vital that we think of

> . . . simple regulations which ensure that society protects the values of natural processes and is itself protected. Conceivably [lands where these processes occur] would provide a source of open space for metropolitan areas . . . Urbanisation proceeds by increasing the density within and extending the periphery, always at the expense of open space . . . This growth is totally unresponsive to natural processes and their values. Optimally, one would wish for two systems within the metropolitan region – one . . . natural processes

preserved in open space, the other . . . urban development. If these were interfused, one could satisfy the provision of open space for the population.

Ian McHarg, *Design with Nature*, 1969

Ian McHarg, and his predecessor, Patrick Geddes, were prophets in an enduring wilderness. Enduring because we have yet to acquire governments with enough political will to think about social and ecological processes inclusively. To bring them to the point of doing so will require, in democracies, bottom-up pressure from the electorate. To bring the electorate to the point of exerting such pressure will require the communication of what's in it for them.

Urban agriculture tends to define itself as a bottom-up, grass roots movement with no time for the top-down elitism of designers. This is misguided. Environmentalism, in whatever guise, demands both top-down and bottom-up initiatives. Freeing up or reclassifying land for urban agriculture requires more than a desire to hold hands and plant vegetables. It requires top-down intervention by planners and local authorities. If urban agriculture is viewed as one of many ways of achieving environmentally productive landscape within, around and outside cities, then those whose business it is to contribute to the design of those cities, their open spaces as well as their built fabric, are vital allies in this project. Urban agriculture in a highly urbanised Western Europe cannot be reproduced in the ways it is being pursued in countries like China, with a much more widespread and direct connection to its traditional farming roots, or even the USA, with its newer immigrant populations from agrarian economies. For urban agriculture in Western Europe to get past the cultivating of one's own city garden, a wider coalition of interest groups needs to be not only tolerated, but welcomed. Anyone with an interest in promoting the more complex, inclusive models of development put forward by Geddes and McHarg, in whatever way, should be able to find a place within discussions of urban agriculture. The environment needs all the help it can get, from as many quarters as it can find.

REFERENCES

Geddes, P. (1968). *Cities in Evolution*. Ernest Benn Ltd, London.

McHarg, I. (1969). *Design with Nature*. Natural History Press, New York.

8

NEW CITIES WITH MORE LIFE: BENEFITS AND OBSTACLES

Joe Howe, André Viljoen and Katrin Bohn

SOCIO-CULTURAL BENEFITS

The literature on urban food growing emphasises its importance in terms of community development and as an agent for urban regeneration, reducing discrimination, tackling crime and generating economic benefit.

The brief overview to follow can be supplemented by making reference to the seminal text by Jac Smit, 'Urban Agriculture: Food, Jobs and Sustainable Cities' published by the UNDP in 1996 and the web sites of the Resource Centre on Urban Agriculture and Forestry (www.ruaf.org) and City Farmer (www.cityfarmer.org).

Urban regeneration

One of the strengths of urban food production, that has been identified in both European and North American literature, is its capacity to make a practical and highly visible difference to people's quality of life (Garnett, 1996, Howe and Wheeler, 1999; Hynes, 1996).

> 'food growing projects can act as a focus for the community to come together, generate a sense of 'can-do', and also help create a sense of local distinctiveness – a sense that each particular place, however ordinary, is unique and has value.'
>
> (Garnett, 1996)

Tackling crime

Hynes (1996) sees tackling crime as one of the prime achievements of the community garden movement in the USA. Often situated in urban areas with very high crime levels, community gardens have been active in rehabilitation work by offering alternatives to drug use, selling drugs and by preventing other criminal activities.

In the UK the authors are aware of schemes in Doncaster where vandalism reportedly stopped once local land was used for orchards and other community activities.

Reducing discrimination

Garnett (1996a) suggests that urban food production provides an excellent means of involving groups who are often discriminated against, such as women, ethnic minorities and the elderly, in sociable, productive activity. Urban food growing has also often provided a valuable means of expression of local or ethnic identity, for example through growing culturally significant produce. A well-known example of this in the UK is the Ashram Acres community garden in Birmingham used by its local, mainly Asian residents.

ECONOMIC BENEFITS

The economic value of urban agriculture cannot be simply compared to the type of finance flow caused by the exchange of money for radishes or apples in supermarkets. Being of small or medium production, preferably organic and of seasonal assortment and aimed at a local market, urban agriculture is a different approach to life and food, competing with or supplementing the growing organic produce in supermarkets.

In the UK there are numerous examples where food growing projects are associated with National Vocational Qualification (NVQ) training courses (Howe and Wheeler, 1999). These range from basic numeracy and literacy courses through to training in subjects like commercial horticulture.

The skills and qualifications gained from this can then be used in seeking employment elsewhere in the horticultural and other sectors.

Urban food growing activities are also valuable educational resources within schools with potential for use in relation to traditional subjects such as science, geography and newer cross-curricular subjects like environmental studies. The 'Growing Food in Cities' report cites a number of case studies where both primary and secondary schools are growing food within their school grounds for precisely these purposes. Education is also a very important part of the activities of most urban farms.

This practical approach to training, if set within the context of a strategy for continuous landscapes within cities, would enhance education and the quality of life for students and citizens by providing a change of environment and heightened sensual experience which is not reliant upon the trappings of consumerism. The observation of outdoor activity and its experience can go a long way to re-establishing a connection with nature. It can introduce a sense of dynamic seasonal change and our part in this temporal environment. Or as Jackson states: 'I believe we attach too much importance to art and architecture in producing an awareness of our belonging to a city or a county when what we actually share is a sense of time' (Jackson, 1994, p. 162).

Producing goods and services

Urban agriculture provides an economic lifeline in many developing countries. In Britain, the commercial aspects of urban food growing have traditionally been inhibited by the general prohibition on selling food from allotments. However, no such restrictions exist in the case of urban farms and some community gardens. Urban farms in particular sometimes derive a significant income from sales of fruit, vegetables and meat through shops, restaurants and direct selling through vegetable box schemes. In addition, urban farms often sell non-food products through outlets such as craft shops and offer services such as horse riding (Howe and Wheeler, 1999).

Supporting local economies

The rise of the food retailing giants and the recent trend towards out-of-town supermarkets in Britain in the last few decades has been spectacular. The decline of the small-scale, local food shop has been equally dramatic. Garnett (1996a) notes that between 1976 and 1987, over 44 000 food retailers closed (31.2 per cent of the overall total) and that by 1988, 90 per cent of all food sales came from just 2 per cent of the stores. The closure of local food stores has left some areas, particularly those in poor urban neighbourhoods, without ready access to food outlets other than high-priced corner shop produce.

This process has led to some startling results; in the USA, a law has had to be passed in order to help feed its poor. Cook and Rodgers (1996) view the combined effects of food industry consolidation and 'red lining', i.e. the practice of supermarkets pulling out of areas where they see insufficient return and an unacceptably high crime risk, as part of a broad pattern in the USA. One in which low income consumers and small farmers have simply been by-passed by the existing network of agri-businesses, supermarkets and food manufacturing conglomerates. This problem prompted the development of a Community Food Security Coalition – linking 125 anti-poverty and sustainable agriculture groups, food banks, small farmers and other organisations. This coalition is dedicated to building alternative systems of food production and distribution that are environmentally sound, socially equitable, health conscious and which provide security through local control. This group has already achieved a notable success in the passing by a conservative congress of the 1996 Community Food Security Act. The act provided funding worth approximately $2 million annually, to qualifying coalition partnerships. This amount might not seem much when compared to a military expenditure amounting to $259.9 billion for 1999 (Stockholm International Peace Research Institute Year Book, 2000), but may none the less be an important first step towards high level recognition of the food security issue.

In Britain, there is nothing yet comparable to the Community Food Security Coalition or Act (see Chapter 6). However the Soil Association, Britain's leading organic charity, has been promoting the closely related concept of 'Local Food Links'. This aims at strengthening local relationships between environmentally friendly food producers and consumers through a variety of mechanisms and outlets. These include local retailers, farm shops, consumer co-operative schemes and vegetable box schemes. Following pilot research in the Bristol area, the Soil Association (subsequently re-named Sustain) engaged in a three-year project aimed at ensuring that local food link schemes exist in every town and city throughout the UK. Sustain aims to lobby local authorities and health authorities, conduct local feasibility studies, launch an advisory service and produce a comprehensive directory of schemes nationwide. There is obvious potential for urban food producers to join local food networks either as part of this formal scheme or informally. Potential players include urban farms, larger community gardens and local authority farms found on the urban fringe. Local authority farms were set up after the First World War to encourage new entrants into agriculture. These farms could supply produce on a significant scale and act as a springboard for new entrants into full-scale sustainable agriculture.

Urban food producers have some obvious commercial advantages over more distantly located producers. Examples include ready access to markets for perishable produce that responds poorly to freezing and conventional storage techniques, and proximity to urban waste products such as waste heat. Imaginative schemes tapping this particular energy source include a tower block in Salford growing food on the building roof and a Sheffield project which plans to use excess heat from a steel works to supply an on-site market garden. This will specialise in higher value exotic vegetables for the local market (see Chapter 9).

HEALTH BENEFITS

Diet

Nutritionists and other health professionals have long recognised deficiencies in diet, particularly amongst the poor. The issue of obesity has risen up the ladder of concern in many developed countries. One in four people in the USA are now classified as obese. In these developed countries diets generally contain excessive amounts of fat and sugar and insufficient vitamin and mineral-rich fresh fruit and vegetables or carbohydrate-rich staples such as bread and potatoes. The effects of this are widely acknowledged, and evident in the excessive rates of coronary heart disease (CHD), obesity, blood pressure and strokes. In response, the UK Government's Health of the Nation report has set dietary targets to reduce the levels of all these problems. The reasons for unhealthy diets amongst the poor may have more to do with a lack of access to retailers selling a good range of affordable fresh produce, than simply to do with pure ignorance. However, food growing in cities can and does help improve people's diets by providing them with access to fresh fruit and vegetables, particularly to those on low incomes, as several cities in the UK now recognise (Howe, 2001). Adherence to organic methods also means that food is usually guaranteed to be free from herbicides and pesticides, although justifiable fears may arise over contaminated land in some inner city locations.

Access to fresh locally grown fruit and vegetables means that people can see where, how and when crops are grown. This is likely to raise awareness about food production techniques and provide knowledge which facilitates a questioning of the advantages of non seasonal imported or processed food. The promotion of CPULs and urban agriculture, in conjunction with healthy eating campaigns, is one way by which diets may be shifted away from excessive fat and sugar consumption in developed countries and one way by which access to food in general is improved in countries suffering from economic hardship.

Exercise

In addition to diet, food growing can provide a useful outlet for increasing the amount of gentle, regular exercise that health professionals argue is necessary to stave off health problems like obesity and CHD.

Mental health

Finally, involvement in gardening or horticulture has also long been recognised for its powers in treating the mentally ill, and recently the Mental Health Foundation in the UK has formally recognised this in their report 'Strategies for Living' (Mental Health Foundation, 2000). A recent study undertaken by the University of Florida (Spence, 1999) is particularly interesting in relation to productive, and continuous landscape which include pedestrian and cycle routes. Its findings suggest that 'just walking around a botanical garden reduces stress levels'. The implication of this is that gardeners and people coming into contact with urban agriculture sites (the public) will all benefit from these health benefits. Financial benefits resulting from this reduced stress can contribute to reductions in public health costs.

As one of the report's authors Professor Jennifer Bradley states,

> the implications for health and well-being are obvious. And for public gardens and the horticultural industry, the implications are good, too. Funding for public gardens is getting harder and harder to come by, so this kind of information gives botanical gardens and arboreta a way to market themselves and more ammunition in seeking funding.

OBSTACLES TO URBAN AGRICULTURE

Having set out its many virtues, what are some of the barriers to urban food production? These fall into three broad categories: regulatory, economic and technical.

Local obstacles include vandalism and theft and a lack of resources, often money and information (see Chapter 9).

Urban agriculture and land-use policy

Urban agriculture is central to the existence of many poorer cities across the globe (Bakker et al., 2000; Ellis and Sunberg, 1998; Smit, 1996). It is only recently, however, that the richer industrial nations of the world and their policy makers have begun to consider the potential benefits of urban agriculture (Garnett, 1996a; Howe and Wheeler, 1999; Hynes, 1996).

With food so high on the political agenda, it is hardly surprising that there has been a worldwide increase in attention given to urban agriculture (Mbiba, 2001). This carries with it implications for land-use policy and regulation. Yet, despite this increased interest in food production and consumption, few studies have examined the nature of recognition and integration of agriculture into regulative frameworks for urban land-use. Within the emerging body of literature on urban agriculture, however, the relationship between urban food growing and land-use regulation has received limited attention, highlighting the relatively uncharted nature of the topic (Howe, 2001; Howe and Wheeler, 1999; Martin and Marsden, 1999; Pothukuchi and Kaufman, 2000).

A recent attempt to review the strategies employed in urban policy to regulate urban food production in different cities around the world demonstrates that the integration of agriculture into land-use policy and city development has remained consistently low (Mbiba and Van Veenhuizen, 2001). In the USA, for example, Pothukuchi and Kaufman (2000) demonstrate a low awareness in land-use planners of increasing urban agriculture activities. Recent studies in Russia and Canada reinforce this bleak picture by demonstrating that many land-use officials do recognise the potential of urban agriculture but find themselves constrained by insufficient budgets (Wekerle, 2001).

The contention here is that land-use implications of urban agriculture deserve further investigation than has hitherto been the case. Now that urban agriculture is considered to be a valid land-use function throughout the world, there is a pressing need to study land-use regulative policy practices of different places to enable lessons of good practice to be debated and considered.

A UK Government Economic and Social Research Council (ESRC) funded survey, led by Joe Howe during 2000–2001, examined the role played by land-use policy in regulating urban agriculture on allotments, community gardens and city farms. All the metropolitan planning authorities in the UK were surveyed, resulting in 32 usable replies, which represents a response rate of 46 per cent. One finding was that 37 per cent of respondents stated that they had experienced conflicts of interest between demands for, and potential changes to, land-use at urban food producing sites. Of these, nine related to development pressure on allotment sites, particularly where these were considered underused. Eighteen authorities had not encountered conflict between urban food and other forms of land-use, suggesting that in the majority of cases urban food production can function harmoniously within an urban area.

The survey indicated that, despite the widespread occurrence of urban food growing activity in the UK, the direct role of land-use regulation policy in relation to urban agriculture is relatively limited. One way in which this issue could be addressed is through the education and training of land-use officials, so that future policy formation will not be undermined by a lack of knowledge.

A number of respondents raised issues as to the level of demand for urban food sites, notably allotment sites, and considered what might be a suitable response on the part of the local authority and its land-use regulation function. Where demand for urban food sites is low, some respondents did identify other potential land-uses that could be adopted for those sites. For example, one authority stated that sites might be used mainly for residential development, since they were primarily located in areas denoted in the development plan as 'primarily residential use'. This answer denotes the potential for conflict which may arise between the need for authorities to utilise brownfield sites in the national quest for protection of greenfield areas, and the need to provide open space and urban food sites.

Financial returns and land

In virtually all European cities, urban food production faces stiff competition from other land-uses such as housing, commerce and industry, which often have a far higher profile and financial return. Overcoming this particular hurdle will significantly influence the development of urban food growing. There is no point in wishing economic obstacles away, but there is good reason to face them objectively. The biggest problem is the prevailing economic system, which only measures direct profits from the development of land. If the prevailing view regarding the comparative value of different activities changed, then planning regulations would follow, and development.

Brownfield versus greenfield

Brownfield sites, those that have had a previous industrial or commercial use, are one of the main sources of development land in existing cities. For CPULs and urban agriculture to become established in existing cities, some brownfield sites will need to be used. For this to happen, the benefits and drawbacks of their use as 'landscape' rather than 'building site' will have to be assessed. It may well be that environmental auditing of different forms of land-use can be used to assess the credits and debits associated with different activities, and so define environmental costs which need to be covered by development. There are now a few tentative moves in this direction in England, for example, the congestion charge in central London (making a charge for private cars entering the city centre) and policies which allow local authorities to take account of the environmental benefits of development and accept a lower price for selling land.

As long as development is concentrated in a few cities, these will almost inevitably need to expand onto greenfield sites, even if their density is increased. If the case for urban agriculture is, as we believe, convincing, then it has a place on brownfield and greenfield sites.

The first step in promoting urban agriculture is to increase the perceived value of development resulting in 'Ecological Intensification' and we hope that making a rational case for the varied benefits of urban agriculture will help to do this.

Technical obstacles

While a great deal of information describing the design of low energy buildings is available, including examples taking account of lifecycle impact due to embodied and operational energy (Viljoen, 1997), little literature is commonly available supporting the case for urban agriculture. Not surprisingly, therefore, built environment professions have traditionally had little to do with food issues. If planning is to be concerned with co-ordinating the use and development of land in the public interest, then the value of urban food growing will need to be publicised far more widely.

Examples of technical obstacles include land contamination, a particular feature of many inner city brownfield sites with long industrial histories. The introduction of organic urban agriculture requires that soils used for growing crops is tested for contamination, to establish if remedial measures need to be taken. Furthermore, hydrological investigations would be required to check that contaminants from adjacent non-organic or polluted sites are not carried by groundwater into organic fields.

Raised beds are often used as a means of limiting the amount of soil which has to be imported into food growing sites, on contaminated soil. The quantities of soil required for extensive urban agriculture sites can be large, and require transport for importation from adjacent areas. Clearly this is a significant factor, which will need careful consideration when determining the viability of new urban agriculture sites. Similar issues exist if roads are to be converted into sites for market gardens; for example, the soil below roads will be heavily compacted by vehicular use, as well as being constructed from gravel and subsoil with all organic soils deliberately removed. Creating fields from roads also usually requires removal of large volumes of asphalt or concrete, but this material can find use as aggregate in general construction. The direct use of road materials to construct raised beds for food growing is not recommended, as leaching of toxic chemicals may occur from the aggregate.

These are some of the technical and environmental difficulties which will have to be addressed by a range of practitioners and professionals to achieve sustainable urban food production. Experiments are underway to 'manufacture' soil from recycled materials, for example, ground glass. The experiments of Phil Craul and much additional detailed information on repairing contaminated soil can be found in J. William Thompson and Kim Sorvig's book, *Sustainable Landscape Construction* (Thompson and Sorvig, 2000). These issues will in certain instances act as limiting factors in the location of urban agriculture fields. A distinction should be drawn between the effect contaminated soil can have on the location of urban agriculture plots and the siting of CPULs. The spaces where CPULs contain circulation routes, parkland and playing fields may be suitable for soils which cannot support edible crops.

The argument that it may be expensive to purchase and prepare urban land for food growing is not sufficient to abandon the case for urban agriculture. After all, just think how expensive it must be for the Netherlands to make the very land on which they then construct towns. Certainly a proactive planning system will be required, allied with a clear presentation of the benefits associated with productive urban landscapes.

REFERENCES

Bakker, N., Dubbeling, M., Guendel, S., Sabel-Koshchella, U. and de Zeeuw, H. (eds) (2000). *Growing Cities, Growing Food: Urban Agriculture* on the Policy Agenda, Die Stifung für Internationale Entwicklung, Feldading.

Cook, H. and Rodgers, A. (1996). Community food security. *Pesticide Campaigner*, **6 (3)** 7–11.

Ellis, F. and Sunberg, J. (1998). Food production, urban areas and policy responses. *World Development*, **26**, 213–225.

Garnett, T. (1996). *Growing Food in Cities*. National Food Alliance, London.

Garnett, T. (1996a). *Harvesting the cities*. Town and Country Planning, **65 (9)**, 264–265.

Howe, J. (2001). Nourishing the city. *Town and Country Planning*, **70 (1)**, 29–31.

Howe, J. and Wheeler, P. (1999). Urban food growing: the experience of two UK cities. *Sustainable Development*, **7 (1)**, 13–25.

Hynes, P. (1996). *A pinch of eden*. Chelsea Green, White River Junction.

Jackson, J. B. (1994). *A sense of place, a sense of time*. Yale University Press.

Martin, R. and Marsden, T. (1999). Food for Urban Spaces: The Development of Urban Food. *International Planning Studies*, **4**, 389–412.

Mental Health Foundation (2000). *Strategies for living*. Research Report.

Mbiba, B. and Van Veenhuizen, R. (2001). The Integration of Urban and Peri-Urban Agriculture into Planning. *Urban Agriculture Magazine*, **4**, 1–4.

Pothukuchi, K. and Kaufman, J. L. (2000). The Food System: A Stranger to the Planning Field. *American Planning Association Journal*, Spring 2000, Vol. **66**, 2, pp. 113–124.

Spence, C. (1999). Botanic Gardens Relieve Stress, UF researchers find. *University of Florida News*. Published at www.napa.ufl.edu/99news/greenspa.htm (accessed 26 August 2003).

Smit, J. (1996). Urban Agriculture: Food, Jobs and Sustainable Cities. *UNDP*, Habitat II Series, Vol. 1, Brussels.

Stockholm International Peace Research Institute Yearbook, 2000, accessed via UK Ministry of Defence web site at: http://www.mod.uk/aboutus/factfiles/budget.htm on 26 August 2003.

Thompson, J. and Sorvig, K. (2000). *Sustainable Landscape Construction*. Island Press, Washington DC.

Viljoen, A. (1997). *Low-Energy Dwellings and their Environmental Impact*. European Directory of Sustainable and Energy Efficient Building. James and James (Science Publishers) Ltd, 47–52.

Wekerle, G. (2001). Planning for Urban Agriculture in Suburban Development in Canada. *Urban Agriculture Magazine*, **4**, 36–37.

9

THE ECONOMICS OF URBAN AND PERI-URBAN AGRICULTURE

James Petts

INTRODUCTION

Urban agriculture is as old as towns and cities themselves, although it has only relatively recently been recognised by national and international bodies for its importance for the sustainability of cities and urban societies. It is an economic activity, engaged in for commercial reasons by an estimated 200 million people, and informally by an additional 600 million people worldwide. The United Nations Development Programmes' (UNDP) groundbreaking book, *Urban Agriculture; food, jobs and sustainable cities* (Smit, 1996), identifies three economic benefits of urban agriculture: employment, income generation and enterprise development; national agriculture sector and urban food supply; and land-use economics.

'Urban agriculture' (UA), for the purposes of this chapter is narrower than the definition used in the rest of this book. In this chapter it refers to small areas, such as verges, allotments, private and community gardens, and balconies within the city, for growing crops and raising poultry and livestock for eggs, meat, milk, etc., for home consumption or sale in neighbourhood markets. 'Peri-urban agriculture' refers to farm units close to town which operate semi or entirely commercial farms and market gardens to grow crops and raise poultry and livestock. The term urban and peri-urban agriculture (UPA) is used to refer to both phenomena.

There are some studies of the economics of urban agriculture, although these are limited, and focus on developing countries or the 'South'. This is not surprising given the number of development agencies and NGOs involved with research, development and facilitation of UPA in the South. UPA is, however, recognised as a global phenomenon and is participated in, studied, and described the world over. UPA in the North (excluding well-known examples such as Russia since the break-up of the Soviet Union) is generally no longer percieved as a response to national crises affecting the entire population, or a 'coping strategy' as it once was during the First and Second World Wars and in previous centuries. In the North it is generally seen as a response to reducing long term environmental degradation or particular local conditions, often related to specific areas of social depravation. In the South it is relied upon by many households for their very existence.

Studies of UPA on a macroeconomic scale, show that the contribution to national economies is not insignificant (See Table 9.1). However, survey data relating to the income earned from urban farming cannot be generalised given the diversity of conditions between seasons, cities, and within cities (Nugent, 2000). Studies calculating UPA's income contribution are unlikely to accurately estimate the quantities of food produced because informal agricultural activity is not usually included. One estimate, however, did calculate that the 30 000 or so allotment holders in London produce nearly as much fruit and vegetables by weight as commercial horticultural enterprises (Garnett, 1999). And one study in the United States (US) of informal gardeners found that the net economic value of 151 plots was between $160 and $178 per year, with a range between $2 to $1134 (The Philadelphia Urban Gardening Project, 1991).

UPA's contribution to employment is also likely to be underestimated because it is not generally recorded in labour statistics of economies, partly because farmers may not count self-employment in UPA as a 'job'. When attempting to calculate the size of the urban agriculture market, difficulties also arise because urban and rural produce is mixed at markets and cannot be determined if the origin is not specified (Petts, 2001). Prices are also difficult

Table 9.1 *Food provided by urban and peri-urban agriculture.*

City/country	Local needs met by UPA (%)	Amount produced annually (tons or litres/day)
Bamako	100 (horticultural)	
Dakar, 1994/1995	70 (vegetables) 65–70 (poultry)	
Harare	small	
Havana, 1998		541 000 (vegetables)
Dar es Salaam, 1999	60 (milk), 90 (vegetables)	
Jakarta, 1999	10 (vegetables), 16 (fruit), 2 (rice)	
Kampala	70 (poultry and eggs)	
Kathmandu	for urban gardeners; 37 (plant foods), 11 (animal)	
La Paz, 1999	30 (vegetables)	
Hubli-Dharwad, 1999	small	40 000 litres/day
London, 1999		8400 (vegetables – commercial) 7460 (vegetables – allotments) 27 (honey)
Lusaka	for squatter population; 33 (total)	
Ho Chi Minh City, 1999	high	217 000 (rice), 214 000 (vegetables), 8700 (poultry), 241 000 (sugar), 27 900 (milk), 4500 (beef)
Hong Kong	45 (vegetables)	
Singapore	80 (poultry), 25 (vegetables)	
Sofia, 1999	48 (milk), 53 (potatoes), 50 (other vegetables)	
Accra, 1999	1 (total)	
Shanghai, 1999	60 (vegetables), 100 (milk), 90 (eggs), 50 (pork, poultry)	

Sources: Smit (1996) and Nugent, R. (2000). *Sustain, City Harvest; the feasibility of growing more food in London.*

to measure due to fluctuations and variations over time and between different markets.

Although urban agriculture currently makes a significant contribution to the food needs of many urban populations, the United Nation's Food and Agriculture Organisation (FAO) has warned that, in the future, the 12 'mega-cities' (10 million plus population) will experience increasing difficulty in feeding themselves (FAO, 1998).

MICRO-ECONOMIC ASPECTS

Models vs reality

In economic utility theory, one model suggests households face choices as to how to allocate their labour and spending in order to maximise their welfare or 'utility' with given resources. It predicts family members will jointly choose how to allocate their labour to maximise income (or spending substitution) over a given period. However, other factors make this analysis more complex when we consider that urban farmers are both suppliers of labour to UPA and consumers of food. Other factors include imperfect labour and land markets, unreliable market information, unclear markets for some inputs, such as credit, gender factors, risk perceptions, and social expectations. The decision to get, or stay, involved with UPA activities leads to changes in how households allocate their time and spending. From a labour supply view, households will produce food themselves if the UPA activity provides a higher return compared with other activities. From a food consumers view, a household will produce its own food when it is less costly (in terms of time and money) than purchasing food (Nugent, 2000).

Within utility theory there are a number of different models determined by the variables used in the function. These can include the income maximisation model where utility is a function of income, the risk aversion model which recognises uncertainty as an important factor, the drudgery aversion model where no labour markets exist, the share-tenancy model where access to land as a productive resource is through non-market mechanisms, and lastly, the farm household model where utility is determined by production and time constraints. All have some validity when considering the different aspects of UPA in reality, but as ever, the reality of UPA turns out to be far more complex than the theory. (Readers are referred to the extensive literature on farm households for further examination of utility models.)

MOTIVES FOR UPA

Surveys of urban farmers' motives for engaging in UPA in the South have been ranked according to their perceived importance (Nugent, 2000). Economic motives of production for home consumption, income enhancement (or expenditure substitution), response to economic crisis, and high prices of market produce are ranked at the top of the list. The ranking is expected to differ compared to farmers in the North, although a number of motives will be shared.

UPA can ensure food security during times of crisis and food scarcity, whether from national emergencies, such as war, or household crises such as sudden unemployment. (Readers are referred to the example of Cuba.) UPA can also enhance food security due to chronic factors. Even in a relatively wealthy country such as the UK, the dramatic rise in the number of 'out-of-town' supermarkets in the 1980s and 1990s has hastened the decline of small, local shops. This decline has left many, particularly the poor and vulnerable, without access to adequate supplies of fresh, nutritious food who instead have to rely more on overpriced, processed foods high in salt, sugar, and saturated fats (Caraher et al., 1998).

Poor families in the South can spend between 50–80 per cent of their income on food and still suffer from food insecurity. Structural adjustment policies of the 1980s and 1990s, leading to the removal of subsidies and price controls, have resulted in rapidly increasing prices of some food commodities (Nugent, 2000). Even with a return to a more stable macroeconomic environment, households may continue to be engaged in UPA, perhaps reflecting an aversion to the risk of food insecurity.

UPA can be carried out by up to two-thirds of urban and peri-urban households, very often informally by women, who combine UPA activities with other activities such as childcare. Urban farmers are not always the poorest residents in the city but are sometimes among those who have lived long enough in an area to secure the means of production, especially land, and have become familiar with the markets for selling surpluses.

Studies in several African countries have found that income from UPA activities makes a significant contribution to the total household income. Drescher (1999) found that home gardens in Lusaka produced an average of three months' income at the average worker's level in 1992, although this was extremely seasonal. In Russia, a survey found that urban farmers in three cities earned an average of 12 per cent of their income from food production in 1995 (Seeth et al., 1998). Nugent (2000) describes a number of factors which will determine the net flow of income of households engaged in UPA activities. These are: farming effort; availability and cost of basic inputs; yields; market access; ability to store; transport; process and preserve; and prices. The *fungible* income of households derived from the substitution of market-bought produce with home grown is a major factor in UPA, as much activity takes place informally.

Employment in UPA provides additional opportunities to the underemployed, temporarily unemployed, or long-term unemployed, where formal employment opportunities are limited. UPA is the second largest employer in Dar es Salaam (UN Centre for Human Settlements, 1992), but relatively few paid jobs exist in UPA beyond the intensive, commercial sector found mainly on the urban fringe (Nugent, 2000a).

Farmers may be engaged in UPA activities because of religious and cultural factors. For example, in Cairo, Muslims will raise small livestock for religious rituals surrounding holidays and funerals.

Even for relatively better off households in the South, the perception of the risk of food insecurity will influence their efforts because of the insurance value of activities. Residents of Hubli-Dharwad keep buffalo as a 'nest egg' to be sold at times of hardship or crisis. At the same time, the buffalo provides the households with a fresh supply of milk, and fuel and fertilizer in the form of dung (Nugent, 2000a).

Food growing can be motivated by factors such as therapy and enjoyment. Recreational gardening and community growing activities are not financially profitable in the North especially when the *opportunity costs* of alternative labour activities are considered. However, as noted earlier, profit is not usually a motive for the informal food grower. In the North, informal participants cite therapy, recreation, exercise, and a household supply of fresh produce as their primary reasons for urban food growing (Petts, 2001).

SUPPLY

Most inputs or *factors of production* for UPA in the South are outside the formal market economy. UPA is an economical use of land for a number of reasons: activities generate income from temporarily available land; techniques need comparatively little space; and some generate a number of jobs. Perhaps the greatest barrier to urban farmers is access to land.

The area of land which is required to make UPA activities commercially viable will depend on a number of factors including: the quality of the land; the use of natural and artificial micro-climates including greenhouses and polytunnels; the type of crops grown; the mix of plants and animals; the price of produce at markets, including produce from rural areas and overseas; levels of other inputs including labour and fertiliser; and the distance of the site to urban markets. One estimate of the land area needed to make an organic box scheme commercially viable in the UK ranges from 1–4 hectares depending upon whether crops are grown in intensive beds or on a field scale (Soil Association, 2001).

Very often land is leased on a temporary basis or obtained informally, known as 'usufruct'. Land used for urban farming generally has a low opportunity cost, which makes it appear to have a low contribution in monetary terms. When the opportunity cost of the land rises sufficiently, the land becomes more desirable for alternative purposes to UPA. This fact can inhibit investment in the agricultural use of land or the use of it altogether, implying that appropriate policies are needed to improve both access to land and security of tenure.

Usufruct increases the total rent that is available from land, providing public and private institutions with an additional source of income. UPA activities may also reduce the maintenance costs for public and private facilities (Smit, 1996).

Critical to UPA and city food enterprises in the North is the scarcity of unskilled labour, shortage of skilled labour, or high wage costs. In some instances this has led to capital-intensive modes of production and mechanisation. In the South, UPA uses relatively little wage labour apart from during seasonal planting and harvesting times. Exceptions to this include Jakarta, Havana, and Shanghai where there is strong institutional support and well-developed commercial sectors (Nugent, 2000a).

Urban farming enterprises on a larger scale may have lower average costs of production compared with smaller enterprises. These *economies of scale* will give the enterprise a cost, and therefore a price/profit advantage within the market place. However, this does not always apply in the urban situation where larger enterprises may face *diseconomies of scale* such as unsellable, large surpluses at some points in the year. Smaller producers, as we note elsewhere, can respond better to changes in consumers' taste, specialise in high value products, and have less need for wage labour, and expensive capital and infrastructure.

Farmer co-operatives can benefit from organisational economies of scale and conduct activities which the individual would find hard to achieve in isolation. Marketing, financing, and technical assistance can be provided through this co-operation. It can also ensure the farmer or enterprise has an almost guaranteed market for the produce within certain quality standards.

The internal economic costs of UPA enterprises include items such as wage labour, plants and seeds, and transport. New UPA enterprises in the North may incur very high initial start-up costs if they are to compete effectively with existing urban and rural enterprises. Cost savings of UPA accruing to other sectors are not generally 'internalised' and may also occur over a number of years, and hence, are likely to be heavily discounted.

Comparative advantage

A comparative advantage exists when either the supply or demand conditions allow UPA to serve the market better by supplying produce otherwise unavailable or prohibitively expensive.

Urban producers can achieve greater efficiencies by using under-utilised resources in the city, such as vacant land, city compost, and unemployed labour. Productivity of UPA can be as much as 15 times greater than that of rural agriculture although yields may suffer from insufficient inputs, skills, and municipal support (FAO, 1998).

As most food is a basic commodity, it has a relatively stable and dependable demand which reduces the associated risks in its production and sale. Urban farmers often have a comparative advantage in speciality produce and markets (e.g. high value and organic) and these specialist UPA activities often become the sole, or a major source of, income.

Farmers close to markets can respond promptly to changes in consumer tastes and demand. Greater responsiveness and availability may often lead to improved quality and nutrient content of produce (Jones, 2001). The close proximity of producers to markets means there is less need for transport, storage and refrigeration equipment and infrastructure. It is also likely to mean better communication, control of supply and quality of produce. For both commercial and home gardeners, the close proximity to market or home saves time and effort and reduces *shoe leather costs*. However, in the North, some urban farmers in a city, located there perhaps for historical reasons, may supply wholesalers and retailers with national distribution systems, so any cost and environmental advantage will be lost.

MARKET ENTRY, DEMAND, AND PRICES

UPA on an informal or small-scale basis, is relatively easy to enter for participants. Farmers can start with a few and inexpensive inputs and limited technical knowledge, probably on land with little or no rent. Productivity is low at this stage but can improve over time with small investments. In the North, market entry on a commercial scale may be hard to achieve due to the cost of land, machinery, labour, and other inputs, which can be significant barriers to entry.

Other barriers to entry to UPA found either in the North or South, or familiar to both, include access to and security of land, relative labour scarcity and undeveloped labour skills, underdevelopment of downstream activities (processing, storage, markets, etc.), and competition from food imports. The cost of transportation of produce to markets can also be a major factor in determining economic viability.

The demand for food is unlikely to change very much with changes in price and other aspects – that is, it has a fairly *inelastic* demand. This would suggest that, even with a downturn in the city economy, urban farmers can still find markets for their goods. However, there may be a high degree of substitutability between foods producing an *elastic* demand for specific products. Hence, with a downturn in the economy, this would suggest the cheaper, perhaps bulkier produce such as potatoes and cereals from more rural districts, will see an increase in demand, whilst the demand for higher value crops grown in urban and peri-urban regions, such as salad vegetables and fresh herbs may be static or fall.

In the South, markets for both urban and rural producers are often well established, although their size and variety of produce vary tremendously with the seasons and they are generally sensitive to many other factors. In the UK, the growth of 'out-of-town' supermarkets has led to a decline in the number of urban markets for city producers. Exceptions to this are the markets which still exist in traditional markets towns, the 500 or so Women's Institute markets which sell home produced goods, and now farmers' markets which have seen a dramatic rise in their number and size in the last few years. This perhaps reflects consumers' desire to support local economies and farmers, and is a reaction to food scares and the globalisation of the food economy.

Prices of produce in urban markets will be determined by the usual interaction of the supply and demand for produce. During the year, there are often fluctuations in price due to the seasonality of supply to markets. Other supply factors affecting price include the relative yields from year to year, the urban infrastructure, and level of agricultural development and support. The determinants of demand influencing price will include the relative price of rural and imported produce, household incomes, and consumer tastes.

MACRO-ECONOMIC ASPECTS

The contribution of a sector to the aggregate or macro-economy is calculated by multiplying the quantity of goods with the market value or price of goods. However, as noted in the introduction, in the case of urban agriculture, official statistics are unlikely to be very accurate because much of the produce is not sold at markets and prices cannot be easily determined. Studies have either estimated total value, the volume of output, or the share of urban food needs produced by UPA. (See Table 9.1.) Although indicative, these estimates are not strictly comparable because they have been produced in different years, using different methodology, and look at different commodities (Nugent, 2000).

One calculation in urban areas of lower income countries found 40–70 per cent of the household budget is spent on food and fuel, with the poorest residents spending between 60–90 per cent of their budgets in this way (Nugent, 2000). This indicates that UPA will often make a significant contribution to a city's aggregate demand or domestic product. There will also be a multiplier effect on urban economies generating output and income, both in related industries, such as tool manufacture, storage and processing, and in completely unrelated industries.

Where UPA and urban rural links are well developed, food prices are likely to be lower because of a reduction in inefficiencies and cost in the supply system, and because of a lower demand due to households substituting market-bought produce with home produced produce. A large number of smaller producers operating on a regional scale will also be expected to have lower food prices compared with a global oligopolies or monopolies.

Well-developed urban–rural links and UPA can be effective buffers to both domestic and external economic 'shocks' as illustrated in the examples of Cuba and Russia in the 1990s. These buffers improve the population's food security and contribute to the resilience and sustainability of the city.

External economic benefits of UPA include cost savings to various sectors including waste management. The ability of UPA to recycle organic waste reduces the municipal authorities' potential costs associated with waste disposal and landfill. Cost savings may accrue to municipal authorities and the private sector through the reduced need for storm water infrastructure and management. Soils will retain water for longer periods, especially when high in organic matter, whilst hard surfaces often result in a rapid run-off of rainwater into drains and catchment channels leading to a greater potential risk of flooding and flood damage. Improvements in air quality brought about by sustainable UPA activities may lead to improvements in the health of the population and to improved labour productivity, resulting in cost savings accruing to individuals, Government health departments and companies.

The external economic costs of some UPA activities can include factors such as pollution abatement and remediation where chemical inputs are used. The cost to the water utilities of cleaning water contaminated with pesticides and other biocides can be substantial. External costs of UPA can also include the costs associated with transport related activities (vehicle emissions/pollution) although these will be significantly lower than those associated with imported produce and air freighted food (Jones, 2001). The internalisation in the market of the external costs and benefits associated with UPA would give sustainable urban and peri-urban produce an additional comparative advantage in markets over similar produce from further afield or from unsustainable production systems.

POLICY CONCLUSIONS

The macro-economic effects of UPA are improved food security and lower food prices, some employment, and contributions to related industries. Although the empirical evidence for these effects is patchy, there is strong anecdotal evidence to support them.

There are demand-led opportunities for urban agriculture and horticulture specialising in niche and other products, such as highly perishable produce, specialised livestock and fish, in cities where UPA is underdeveloped (Garnett, 1999). The demand-led opportunities need to be matched, however, by supply-led growth in more deprived areas, to encourage equity of access and affordability between different social groups.

Local food purchasing policies in the statutory sector (hospitals, prisons, schools, etc.) could provide a major impetus to the development of UPA given that a large proportion of all food consumed is through this sector. Price and/or quality advantages would be needed to enable justification by the relevant authorities.

The provision and reasonable pricing of land and water resources is essential to develop UPA in the city. Within a supportive framework, policies are needed to protect and improve the security of land tenure for urban farmers. In the UK, this could take the form of amending planning policy guidance (PPG) relating to allotments, fringe farms, and other urban spaces used for food production.

Planning regulations and guidance should take into account the needs of urban farmers to invest in refurbishing facilities, or in new ones, to develop processing, storage, and packaging facilities, thus enabling them to 'add value' to their products and improve their turnover and incomes.

As with other industries, UPA is dependent upon the city's infrastructure including the transport system, public utilities such as water management systems, and other services including labour and produce markets. Such infrastructure, if it does not exist, may be expensive to provide and maintain. Policies should be developed to encourage small-scale, inexpensive systems designed to address the most critical weaknesses in a city's capital. With relatively little investment, such systems can greatly increase urban farmers' productivity and viability.

Regional, national, and international policy changes are needed to overcome structural barriers and distorted markets in the urban food supply system. Potential new entrants operating sustainable modes of production would be encouraged by the internalisation of the external economic costs and benefits. Appropriate standards, incentives, subsidy, taxes, and regulations could be used to internalise external

PLANNING FOR CPULs: URBAN AGRICULTURE

Figure 9.1 *An allotment alongside a railway viaduct in London. Land adjacent to railways often provides a continous network of open space, which could be connected into CPULs.*

costs and benefits and shift profitability in favour of sustainable practices and encourage a thriving local food industry within and around towns and cities. For example, a tax on aviation fuel would help reflect its real economic cost to society and give produce grown closer to home an advantage, whilst a subsidy for organic production would reduce the external economic costs incurred by water companies and their customers, caused by having to cleanse water of pesticides and artificial fertilisers.

Case Study 9.1 *City Harvest project in London, UK*

Sustain, an alliance of a number of public interest organisations in the UK, initiated the City Harvest project in 1998 to examine the nature of UPA in London, and to encourage its residents to grow food. The project followed from the Growing Food in Cities project which looked at the benefits of urban agriculture in the UK. City Harvest organised a local food festival to celebrate the range, quality, and diversity of food grown in or near London and researched the existing aspects of urban agriculture, and its potential to contribute to the food security and sustainability of London in the future.

The study found that London's 'ecological footprint' is estimated to be 125 times the capital's surface area with food accounting for around 40 per cent of this. London's residents, visitors and workers, consume 2.4 million tonnes of food and produce 883 000 tonnes of organic waste per year. The food industry makes a significant contribution to London's GDP (about $122 billion) with around 11 per cent of all jobs found in the food sector, although most of the food consumed comes from rural districts and other countries.

Almost 10 per cent of Greater London's area is farm land, of which 500 hectares are under fruit and vegetable cultivation, and there are around 30 000 active allotment holders and an estimated 1000 beekeepers. Overall, the area under commercial cultivation is continuing to decline due to development pressures, labour scarcity/high wage costs, a skewed system of Common Agricultural Policy subsidies, competition from imports, and planning restrictions. The main area for commercial cultivation is the Lea Valley region in North East London.

Figure 9.1

Figure 9.2 *A typical allotment site in England.*

Figure 9.2

The Lea Valley region typifies a declining industrialised horticulture. This once thriving area for food production has shrunk since the war and the industry now only covers an area of 120 hectares under glass. It has a high productivity, with 200 or so horticultural enterprises ranging in size from less than an acre to 20 acres, with production often automated and hydroponic. The enterprises sell produce nationally to wholesalers and supermarkets.

The remnants of commercial urban agriculture in London, such as the Lea Valley example, provide an opportunity to redevelop, modify, and diversify the industry towards a more sustainable system. Conversion to organic, development of sustainable, social enterprises, and production for local London markets, such as farmer's markets, restaurants, and co-ops, would utilise the existing infrastructure and change the mode of production. Barriers would have to be overcome to enable this, such as encouraging organic horticultural production through the CAP and government programmes, changing planning policy to discourage inappropriate developments on the urban fringe whilst allowing fragmentation of farms to create smaller, more diverse holdings, and internalising the external costs so food prices reflect their true cost and make the sale of produce at local markets preferable.

Figure 9.3 *A vegetable market in Burkina Faso.*

Case Study 9.2 *Phd thesis on the economic costs and impact of home gardening in Ouagadougou, Burkina Faso*

Sibylle Gerstle of Basel University, Switzerland, studied the economic costs and the effects of home gardening in Ouagadougou, Burkina Faso. Completed in May 2001, Gerstle looked at both a whole population group and, as samples, home gardeners on three sites with differing social and economic structures. Within this, Gerstle also considered the possible links between the economic dimension of home gardening and the health status of home gardeners.

Burkina Faso is one of the poorest nations in the world with a Gross National Product per capita of only $240 and with an infant mortality rate of 210 per 1000 for the under fives. Burkina Faso's capital, Ouagadougou, has a population of 750 000 and an average growth rate of 6.8 per cent. It has 48 home gardening sites varying in size and from season to season.

One result from Gerstle's study was that in both the dry and rainy seasons, home gardeners had a lower average monthly income than their counterparts practicing any activity other than home gardening. This finding is unusual because other studies in Africa (Zalle 1993 (Bamako), Sawio 1993 (Dar es Salaam), Egziabher 1994 (Addis Ababa), and Mougeot 1999 (Lome)) have found that urban farmers have a higher than average income. However, an indirect estimate for income calculated on the basis of prices obtained for vegetables, was higher than the directly estimated one for all three sites. Additionally, in all three sites, the average monthly income of households was higher in the dry season than the rainy season, and income varied considerably between sites due to factors including the quality of irrigation water and the crop patterns.

In the whole population group, more than half of the predictable expenditure was on food. In the rainy season home gardening households had considerably lower expenditure on food than non-home gardening households, whilst in the dry season, home gardening households spent an equivalent amount on food as non-home gardening households. Hence, the contribution of home gardening to household food security and 'fungible' incomes was mainly confined to the rainy season, reflecting the importance of the

Figure 9.3

seasons and climatic factors to urban gardeners in Ouagadougou.

With respect to the health status of the population, the most common aliments were malaria, gastro-intestinal diseases and respiratory diseases. The study found that the average number of days of illness was equivalent in both home gardening and non-home gardening households, and, surprisingly, there was no correlation between the health status of the population and their socio-economic status. Hence, the expected health risk related to home gardening was not found and the lower monthly expenditure, lower income, and a lower expenditure coverage rate were not directly related to the health status of home gardeners.

REFERENCES

Blair, D., Giesecke, C. and Sherman, S. (1991). The Philadelphia Urban Gardening Project, *Journal of Nutrition Education*, **23**, 161–167.

Caraher, M., Dixon, P., Lang, T. and Carr-Hill, R. (1998). Barriers to accessing healthy foods: differentials by gender, social class, income and mode of transport. *Health Education Journal*, **57**(3), 191–201.

Drescher (1999). *Urban agriculture in Lusaka: case study*. IDRC.

FAO (1998). *Feeding the cities, Sofa report*. FAO.

Garnett, T. (1999). *City Harvest the feasibility of growing more food in London*. Sustain.

Jones, A. (2001). *Eating oil: food supply in a changing climate*. Sustain.

Nugent, R. (2000). The impact of urban agriculture on the household and local economies. In *Growing Cities, Growing Food*, Deutsche Stiftung für internationale Entwicklung (DSE).

Nugent, R. (2000a). *Urban and Peri-Urban Agriculture, Household Food Security and Nutrition*. Discussion paper for FAO-ETC/RUAF electronic conference 'Urban and Peri-Urban Agriculture on the Policy Agenda'.

Petts, J. (2001). *Economic costs and benefits of UA in East London*, Sustain.

Seeth, H., Chachnov, S., Surinov, A. and Von Braun, J. (1998). *Russian poverty; muddling through economic transition with garden plots*. World Development No. 26.

Smit, J. (1996). *Urban Agriculture; Food, Jobs, and Sustainable Cities*. UNDP.

Soil Association (2001). *A Share in the Harvest*, Soil Association.

UN Centre for Human Settlements (1992). *Sustainable Cities Programme, Dar es Salaam: Environmental Profile*.

10

CHANGING CONSUMER BEHAVIOUR: THE ROLE OF FARMERS' MARKETS

Nina Planck

First, a bit of background about London Farmers' Markets. I grew up on a farm in Virginia and I've sold at farmers' markets all my life. My parents still make a living exclusively from farmers' markets. They grow fruit and vegetables and sell them in and around Washington, DC.

The lesson for us when we first started farming in 1979 – and weren't doing very well standing by the side of the road in little town near our farm – was that we had to go into Washington, meet with other farmers, and sell at farmers' markets. You have to go where the people are. Now this is the beauty of urban agriculture – the people are already there. But my background is working with farmers to get them to travel *into* town, to benefit from the presence of *other* farmers, and also from the large number of people in a small space. In four hours, many more people walk by the farmers' market than by the farm gate. So my interest is in increasing income to farmers, getting fresh food to people in cities, and having that work commercially for farmers.

The rules at LFM markets are: every producer has to come from within a hundred miles of the M25 – or rather all the food is produced within a hundred miles; you have to grow or produce it yourself; and you have to be at the market to sell it. Many farmers' markets in the UK have a smaller catchment area than we at LFM have. They require that all the farmers come from 20 or 30, or sometimes as far as 50 miles away from the market. We didn't think that made sense for the regional farmland that is required to feed a population the size of London's. Nor did it make sense to go looking for farms which didn't exist in the suburbs of London.

In 2002, we ran twelve weekly markets. We'd like to do more with these markets, and one goal is to have more local foods. I mean local in two senses: first, that we include more farmers who are closer to the market. At the moment the vast majority of our farmers grow their food *less than* 50 miles away from the market, so we're pretty close to the catchment area of a lot of the rural farmers' markets in this country. Secondly, more local foods would also mean tightening up LFM's few exceptions to the local-ingredients rule for processed foods. For example, we allow bakers – and probably always will – to buy flour. Bakers are the only producers not required to use ingredients grown within 100 miles. However, there are lots of arable farms struggling in the commodities market in this country. We're good at growing cereals. The EU export market for cereals is essentially phoney, because it is heavily subsidised. Therefore, if we want arable farmers to be producing something economically viable, rather than just paying them to plant hedgerows, they ought to be looking at different markets. I think there's a big market out there for muesli from local ingredients, and for baked goods made from local or regional flour.

More organic food is another goal. About three years ago, when I first started talking about farmers' markets with people here, there was strong support from two sectors in particular: local food activists and the organic world. That was a welcome thing, because it meant there was lots of political support for farmers' markets. However, I didn't think that we should sacrifice certain farmers on the cross of ideological purity on organic food or local food miles. That is partly why I set the LFM catchment quite wide, and also why I don't believe farmers' markets should be for organic farmers exclusively. All farmers should have a chance to meet the public, so that they can learn what the public want, and also so they have chance to earn the retail price. A third reason we shouldn't exclude conventional farmers is this: if they do *anything* to change methods of production when they sell direct, it will be in the direction of greener agriculture, not the other way round. You don't meet consumers who say, 'I quite like your carrots, but I wish you'd use more organophosphates.'

A fourth reason for including non-organic farmers is that there are many green practices in agriculture that fall short of legal definitions of organic. I would like farmers to find outlets for those products, and farmers' markets are a good example. My own family started out using scarcely any, but a few, chemicals. Then we stopped using chemicals entirely, but we've never chosen to certify our farm because we use a seawater-based fertiliser that Virginia certification standards – and now US certification standards – don't allow. We prefer to keep it. We think it gives us healthy plants and tasty vegetables, and that it's good for us as farmers and consumers. We didn't think certification was right for us. Yet we get premium prices at all our markets because our relationship with the consumer is direct, and they are satisfied that we grow our food the way we say we grow it.

I think there is room to let a thousand green flowers bloom. I'm happy to say that a lot of LFM's farmers are moving in a greener direction. 'NO INSECTICIDES ON OUR FRUIT' is a much more common sign than before. Remember, farmers have lots of reasons to go green. Some are concerned about their own health and spraying. Some are concerned about getting better prices. Some are concerned about the environment, others about consumer health. In the farmers' market all these motivations come into play in different quantities with different farmers and different consumers.

We would also like to see more viable city farms. There are lots of city farms, and they own valuable land. But too often, in my view, they're just educational. I'd love to see those farms have a product to sell to local people. I do know that Dean City Farm in Merton, near our market in Wimbledon Park, has five acres of arable land, and not a thing growing on it. That's pretty valuable real estate. I understand that this city farm has to raise £5000 a month, but it hasn't any product to sell to raise that money. I also know that on a plot this big, if you're growing commercially, you could be making – assuming just two markets a week – £1000 a week. That's a modest estimate.

Let me show you how little land you need. A friend of ours bought a farm in Athens, Georgia. The first year he planted half an acre, set up one box scheme, and attended one weekly farmers' market. He employed one person, part-time. His season was March to November, much shorter than in the UK. His gross sales in the first summer, on one half an acre, were $20 000, all of which he reinvested in the farm. (He worked as a carpenter in the winter.) In the second season – which was two months shorter, for non-farm reasons – he grew on one acre and grossed $30 000, with a profit of $10 000. The crops Andrew grew – cut flowers, basil, rocket, French beans, sweet peppers, and other vegetables – are low-volume but high-value. We'd like city farms to grow these things.

Another goal is to include more ethnic consumers, producers, and foods at LFM markets. All the foods would be seasonal and regional, of course. This is well developed in the USA, in most cases naturally. There wasn't a central government policy to get Hmong and Laotian producers in Massachusetts or Wisconsin to grow stuff for the customers coming to city farmers' markets. It just happened – first, because 'ethnic minority' consumers came to farmers' markets, many of which are in cities, and started to ask for certain things, and then the 'ethnic majority' farmers started growing what they wanted them to grow, and then new ethnic minority producers said: 'Why don't *we* grow these things?' There have been more sophisticated efforts to encourage this. But mostly it just happened because it's a market place, and the market has a magic quality of getting two people together to trade.

It's happening at farmers' markets all over the USA: there are Mexican growers and consumers in California and Texas, Asians in the Northeast, black Americans in Washington DC. At one of our DC markets, where the customers were mostly black, we learned what they liked to eat, and how. We learned about collard greens and other leafy brassicas like mustard, which are traditional Southern foods, and we learned that many people preferred to eat them after one frost had made them sweet.

In London, there's a Bengali women's co-op in the East End. At the moment they grow, like most interesting women's projects, not even as much as they can eat themselves. So I say: just grow a little more, because we're going to have a farmers' market in Tower Hamlets (opened in April 2002), and the neighbourhood is some 70 per cent Bengali.

Finally, we'd like to see more use of derelict land and buildings. I've alluded to this on city farms – empty, disused space drives me nuts. In pictures of the Havana growing projects, you see tiny spaces filled with vegetables. In London there is a lot of space. It's the least dense city in Europe. Our tradition has everyone with his own little green patch behind his house. But it's not only gardens I have in mind. It's public space in corners and places like that. It would be fun to use the Barcelona model for city parks. I haven't been there, but I understand that instead of big green spaces, they built tiny ones; little urban oases, where people could sit down. It was a new model for open urban spaces. I'd like to see the same approach for growing spaces.

All these goals have one thing in common. In London, the string linking these goals and the *advantage* for each of them is that there is already a known system of selling local foods. It's not only our markets; LFM organises ten weekly markets, but there are others, for example, the high-quality food market at Borough, in Southwark. The forums for selling local food already exist. I've often seen local growing projects founder on having nowhere for the produce to go. They're either growing just for home consumption, and have too little to sell, or they've got no funding to carry on, that is, no income. Selling retail is the answer.

We would be very pleased to allow allotment gardeners to form co-ops and sell at LFM's. Furthermore, an allotment stall at the market would be a great way for local authorities to meet the statutory Policy and Planning Guidance 17 requirement that local authorities publicise and encourage the use of allotments.

Demand for local food is high. Demand is everywhere. We can't meet it. We don't have the farmers. We spend about a third of our time recruiting farmers. We're working on that. What we don't need to work on is finding customers. The customers are there. Any local growing project seeking an outlet has one.

11

THE SOCIAL ROLE OF COMMUNITY FARMS AND GARDENS IN THE CITY

Jeremy Iles

Do we have a form of Urban Agriculture in the UK? If compared to the examples from Cuba or some African countries the answer must be no. But we do have a thriving network of community gardens, city farms, school farms, allotments and community run growing projects, all of which constitute urban food growing initiatives which, in total, engage something approaching 10 per cent of the population in some way[1].

That the urban area could be used for productive food growing was demonstrated in the Second World War, when large areas of urban parks, gardens and recreational areas were turned over to food growing. However, since the 1950s 'home grown' food has declined in importance as the power of supermarkets and the global transportation of food has increased.

WHAT IS A COMMUNITY GARDEN OR A CITY FARM?

To start with we will define a community garden and a city farm. These are local projects managed for, and by, local community groups. They are sometimes run as partnerships with local authorities, but the essential feature is strong local involvement. They exist mostly in built-up areas, where their creation has been a response to the local community's lack of access to open, informal, community-run green space.

City farms are also known as urban farms, children's farms, or community farms. Allotments are not generally community-managed, but there is a growing movement for allotment groups to take on devolved management from the Local Authority which will move them from the 'statutory' sector (albeit with legal protection), and more towards the community-led sector. And within the allotment movement there is also a growing number of groups that are consciously setting up community managed working with innovative schemes to bring in more community support.

There is no 'typical' project, as each develops according to the local area and in response to the developing needs of the local community. These projects deliver a wide range of community-managed services in response to local needs.

All community garden and city farm projects are run by a local management group and all have strong volunteer involvement. They are places where people of all ages and from all sections of the community are welcome.

Most projects provide food-growing activities, training courses, school visits, community allotments and community businesses. In addition, some provide play facilities and sports facilities, and after school and holiday schemes.

THE BENEFITS OF A COMMUNITY GARDEN OR CITY FARM PROJECT

Community gardens and city farms are extremely flexible and adapt to the changing demands of the local community. What they have in common is encouragement of social participation and the creation of sustainable communities, through:

- the creation and management of community-managed green space.
- individual development – by giving people opportunities to take part in many situations, from learning to grow their own food and meeting people, to undertaking a management role.
- social inclusion and cohesion – by being accessible to people from many different backgrounds and

offering people who are isolated ways to extend their networks.
- strengthening and regenerating urban and other communities – by enabling people to get involved in community activities. Such involvement fosters pride and identification within a local community, which encourages people to participate in public affairs.

Through social participation, community gardens and city farms can help people learn new skills and gain self-confidence. Access is free which reinforces their integrative potential.

Bradford City Farm

A project working with young people, especially from the Asian community, showing the benefits of farming and growing.

'You don't have to be sitting behind a desk,' says Manager, Rob Dark, 'to learn something that'll prove useful in later life! We do a lot of work with young people, including those excluded from school, and have developed a successful "work-readiness" programme for final year students. There are many links between farming and growing and the National Curriculum, and local schools are regular visitors.'

The project also provides placements for long-term unemployed people and adults with special needs; helps with community planting and garden design; and sells produce at the local farmers' market.

Heeley City Farm, Sheffield

A project in a deprived inner city area employing 34 staff, supporting up to 100 volunteers, and welcoming around 100 000 adult visitors a year.

'The driving force here,' says David Gray of Heeley City Farm, 'is our commitment to supporting those most in need in our community. We offer day care and training for people with learning disabilities, and NVQs in horticulture, agriculture, basic skills and English as another language. And each year we provide environmental, food and health-related education to 5000 school children.'

'I think our greatest achievement is employment of local people. 83 per cent of our staff were formerly unemployed and 60 per cent of them live within one mile of the farm.'

The project is also a founder member of Heeley Development Trust and Sheffield Environmental Training, which together employ 60 people. This led on to their involvement in the creation of the new Heeley Millennium Park.

New projects in the pipeline include devoting 20 hectares to growing food for the local economy, and a partnership with Sheffield Black Community Forum, working with groups on local issues and involving ethnic minority groups in environmental projects.

Projects contribute directly to community development by generating social participation, and promoting urban regeneration through providing:

- additional green space in the urban environment
- a 'gateway' into both informal and formal educational opportunities
- local education about food growing and caring for animals
- adult education in a wide range of subjects, e.g. gardening, horticulture, animal husbandry, English as a second language, computer skills
- school visits and educational activities
- pre school activities

- play facilities and sports facilities
- after school and holiday play schemes
- placements for people with learning difficulties and other special needs
- community allotments and community orchards
- community enterprise development and training, e.g. cafes, horse riding lessons, garden centre activities, community businesses.

They appeal to a wide range of people and can create opportunities for disadvantaged and minority groups. They can also generate local economic activity and community businesses.

Community activity is the hallmark of city farming and community gardening projects, and it is fundamental to promoting well functioning and sustainable communities. This in turn will lead to the development of a greater confidence and skills base within a local community – social participation leads directly to the growth of social capital, that is, the ability of a community to take an interest in and to shape its own future.

Welbeck Road Allotments Association Trust, Derbyshire

Derelict allotments have become a showcase for urban regeneration.

'It all started with a hedge!' says Manager, Mike Gosnell. 'Plotholders were demoralised by the activities of thieves and petty vandals, and decided that a hedge might keep troublemakers out. We applied for Local Agenda 21 funding, employed a hedge-layer, and enlisted local children to help. The Council was very impressed with the results.'

Now they have a hazel coppice, a clubhouse, and a wildlife pond where a tip used to be. They are planting an orchard, setting up a community polytunnel and planning many other projects.

They have won local awards, and have received much publicity.

'From feeling powerless and dependent on the council, we've become proactive. We initiate ideas and get funding from a number of sources. We've transformed the threat of vandalism into the regeneration of the site.'

POLICY ON FOOD PRODUCTION AND SUSTAINABILITY

If social participation in community-based gardening and farming activities can lead to a greater understanding by the urban majority about the food produced, largely in rural areas, then this may contribute to a greater common understanding, and thus social cohesion. In turn, the rural farming population needs to acknowledge that efforts to create educational opportunities around food growing and environmental issues will bring new demands for more responsible agricultural practices and environmental stewardship.

The government has recognised this, and through work at the Department for Education and Skills, has piloted a new 'Growing Schools' initiative, to promote active learning opportunities through growing and farming. City farms and community gardens, allotments and school growing areas will all have a role to play. By laying the seeds of a better understanding of food production at schools, the hope is that a new generation will grow up with a better understanding of the true value of food, and with the knowledge to make informed decisions about future food policy.

In addition, the Home Office is keen to promote a more socially inclusive society, and the Department for Transport and Local Government has recognised the value of promoting increased social participation through a growth in the number of community-led gardening and farming projects. At a policy level we can expect to see policies being introduced which in turn might lead to a growth in the number and geographical spread of community gardens and city farms, as well as school growing projects and community-led allotments groups. The provision of resources will, needless to say, be another matter.

ENCOURAGING AN EXISTING OR NEW PROJECT

Whilst local authorities might be tempted to set up a 'community project' from top down, it is better to facilitate and encourage genuine grassroots social participation and community activity. Awareness, and promotion, of the benefits through the local community will reveal community or school groups who have, or would like to, set up new projects.

Local authority officers and councillors will need to develop supportive relationships with these groups to examine ways of making the projects stronger, well managed and financially stable.

Culpeper Community Garden, London

An organic inner city community garden, worked by individuals and by groups.

'We hear a lot about community involvement and empowerment: in practice it can be the most uplifting thing,' says Clare Sutton, Project Worker at Culpeper, and Chair of the Federation of City Farms and Community Gardens.

The 48 plots on the community garden are a real local resource: used by a playgroup, a school, a mental health day service, and an organisation for people with learning difficulties.

The garden is open to the public and is popular with local workers at lunchtime. Garden workers and volunteers help with maintenance. Events such as a children's summer art project, pensioners' strawberry teas and plant sales are organised to bring local people into the garden.

How are projects like this funded?

Projects are funded from a variety of sources depending on their stage of development and sophistication.

Local authorities often provide core funds for salaries, and there are several streams of Lottery funds, which might be available for feasibility studies and project development. In addition, projects can sometimes access regional development funds, single regeneration funds and European funds. When applying for funds, help in kind and volunteer time should be valued as match funding.

Once established, many projects become more self-sufficient by charging fees for training courses and placements, and developing income generating activities such as cafes, horse riding, sports facilities, after school and holiday clubs. This kind of service provision is also often grant-aided by the local authority. However, when successful, such social enterprise becomes not only an income generator, but also another kind of social participation, leading to further strengthening fo the local community. Projects also fundraise from corporate and charitable sources, as well as from the local community.

What resources are needed to run a successful community-led project?

The greatest need that projects have, especially when they are starting out, is local political support through committee and officer time: a clear sign that the project is valued and welcomed. Projects need clear support mechanisms through the council to develop a genuine partnership. This will help greatly with forward planning, credibility, and identifying funding needs.

Local authorities can also help by identifying land and making it available for a new project at a low rent. Needless to say, projects may also need financial assistance, particularly to help in the longer-term stability of a project.

Although difficult to quantify, local authorities should recognise the social participation and 'community benefits' that these projects bring: it is not simply a question of profit and loss, and subsidies may be seen to bring benefits far in excess of the financial grant.

Wellgate Community Farm, Romford

A small suburban project, this is a resource for everyone in the local community.

'It's all about including people,' says Manager, Rob Gayler, 'providing a "growing space" for groups and individuals. People come to the farm because they want to work with animals, but they stay because of the people. Visitors are welcome but the emphasis is on involvement, and we have volunteers of all ages.'

The project offers training for adults with special needs and has a horticultural therapy team for people recovering from mental illness. They work with youngsters who are non-attenders or excluded from school, and many of them go on to further training and employment.

Schools are encouraged to visit the farm, and the farm visits local schools! 'Sometimes it's easier to move the animals than the children,' *says Rob, 'so we take them in the horsebox and meet the children on the school field!'*

A VISION FOR THE FUTURE

The value of social participation through city farms, community gardens and other related community-led growing projects has never been better understood by policy makers. We can therefore look forward to a gradual growth in their provision, and hopefully in a recognition that they provide a valuable local service which needs proper capital and revenue funding. If every school child could be introduced to the concepts of sustainable food growing, and every town has access to a city farm or community garden, we would be getting somewhere. That vision may be some way off, but it is a vision we will work towards.

THE FEDERATION OF CITY FARMS AND COMMUNITY GARDENS

The Federation of City Farms and Community Gardens (FCFCG) is a UK-wide charity representing city farms, community gardens, a network of school farms, some community-led allotment groups, and community groups involved in projects in parks. In total, we represent 65 city farms, more than 1000 community gardens, 75 school farms and around 20 allotment groups. In addition, we have records of some 200 potential farm or garden projects.

Between them, our member projects employ 550 staff and engage and empower thousands of volunteers, with two thirds of projects run entirely by volunteers. Projects attract over three million visitors and regular users each year and have a combined turnover of around £6 million.

PLANNING FOR CPULs: URBAN AGRICULTURE

The Federation of City Farms and Community Gardens provides:

- advice and information packs on developing community gardens and city farms
- animal welfare, health and safety, and child protection guidelines
- model documentation for establishing a new group or charity
- case studies on good practice
- individual advice to new projects with project visits from experienced practitioners
- advice on sources of funds and how to apply.

In addition, FCFCG may be contracted to undertake feasibility studies and to work with local groups to develop funding bids.

NOTE

1. Number of visitors per year is based upon the following estimate:

   350 000 allotment plots × 2 people average
   700 000

   65 city farms @ 50 000 visitors per year
   3 250 000

   75 school farms @ 1000 pupils

   75 000

   1000+ community gardens @ 100 people per year 1 000 000

   Total estimate   5 025 000

12

RECYCLING SYSTEMS AT THE URBAN SCALE

Dr Margi Lennartsson

The Henry Doubleday Research Association (HDRA) is an organisation that deals with organic horticulture in its widest sense – from domestic gardening to allotments, landscaping, and commercial organic production. One of our key concerns is the composting of organic waste for use in urban horticulture. Over the past decade we've seen the dawn of a new era for waste management. Gone are the days when we would think it acceptable to collect our waste and then just dump it in a big landfill site, or even pump it into the sea. Instead, we've seen recycling plants, including composting facilities, established all over the country.

The first big step towards a greater emphasis on composting came in 1990 with the publication of the government White Paper 'This Common Inheritance,' which set very clear targets for recycling. Over the years these targets have been updated and expanded. A second milestone came in 1999, with the introduction of the EU Landfill Directive, which sets clear limits on how much material can be sent to landfill, with targets based on landfill volumes in 1995. Thus, by 2010 the UK must landfill no more than 75 per cent of the volume of waste disposed of through that route in 1995, 50 per cent by 2030, and 35 per cent by 2020. This represents a dramatic shift in our reliance upon landfill. The EU Landfill Directive also refers specifically to biodegradable waste, with the implication for the UK that by 2010 we will have to divert three and a half million tons of biodegradable waste per annum from landfill. At the moment we divert about one million tons, so although we already have seventy big composting sites across the country, this is really is just the beginning: we still have a long way to go before we can be confident of achieving the targets.

There are various ways of disposing of biodegradable waste which avoid one of the drawbacks of landfill – the consequent free emissions of methane, a very potent greenhouse gas. One option is anaerobic digestion, which can generate some usable energy, but this method still leaves a residue which has to be disposed of somehow. Incineration is another well-supported option, but an imperfect one, since wet biodegradable waste is not particularly fitted to incineration, as it reduces the calorific value of the waste. Thus, the third option, composting, looks to many (including HDRA) as the most cost-effective way of moving forward.

Actually, there are a number of different composting options from which to choose. There are on-site options, including home composting, (i.e. composting material in our own domestic gardens), community composting, and on-farm composting. And then there are centralised composting units, either open-air systems, which are the most common at present, or the in-vessel systems that are beginning to emerge. Let's examine these options in detail.

Many people, and especially urban horticulturalists, are very familiar with domestic on-site composting in the garden, in which biodegradable waste is collected from the garden and placed on a compost heap along with selected kitchen vegetable waste. Most local authorities are involved in active initiatives to promote this sort of composting, because it's obviously one of the best ways of diverting biodegradable waste. No transport at all is involved, since the waste stays at home, and households create their own compost, which can then replace some of the inputs that would otherwise be used in the garden. Local authorities are also getting a lot of help in encouraging more domestic composting, for example through the efforts of HDRA and other organisations in promoting composting education programmes: how to set them up, how to get people involved, and how to present the message. There is also the Master Composting Manual, a guide for local authorities on successful implementation of home composting schemes. There are good functioning examples of campaigns to emulate, such as Rugby Borough Council's 'Rugby Rotter' and the Coventry Composter Newsletter. And there are large amounts of practical information available to the public on home composting, such as that produced at HDRA.

An alternative centralised option is open-air windrow composting. Not everyone can compost their own domestic waste at home. There are always going to be some wastes that have to be collected at the curbside and brought to central sites for composting. At present these facilities are usually located outside the city boundaries, or even in the rural areas. The material is fed through a large shredder in which it is macerated to make it uniform and more accessible for micro-organisms. These facilities do not have to be large-scale, although many are. Smaller-scale equipment is available for use in parks and by garden landscapers, with the shredder simply mounted on the back of a tractor.

After shredding, the material is placed in what are called windrows – long heaps, usually around 2 metres high and 3 metres wide. The material stays in the windrows for about twelve to sixteen weeks, and it is during this time that the composting takes place. The material has to be turned regularly, because composting is an aerobic process and requires air. When the material is macerated, the micro-organisms within it start to break it down, generating heat – composting is basically decomposition under elevated temperatures. If the material were just left undisturbed the temperature could easily climb to 70 or even 80 degrees centigrade very quickly. But the most effective temperature for the composting process is around 45 degrees, so to maintain that temperature the material is turned regularly, around once a week, sometimes by dedicated equipment that straddles the compost heap, turning it as it moves down the row.

After the composting process has been completed, the material is normally moved into an even bigger pile, where it remains for another month or two, or perhaps even longer, as it goes through its maturation phase. After that, it is passed through a large sieve or screen, which sorts it into different size fractions, with the big material taken out, and the woody stuff going back for further composting. And thus, almost by magic, at the end of this process you have wonderful, humus-rich compost.

Within urban settings, the in-vessel system may be more appropriate. Here, the composting process takes place indoors, mainly to control the emissions. With a well-run outdoor system you can control odours quite well, but there are times when any collection of waste will give off odours, and that's obviously unacceptable within an urban area. Hence in Holland, for example, with its very high population density, in-vessel systems have been developed and deployed much more extensively.

When you compost indoors, instead of turning the material, you normally force or suck air through it, to keep it aerated. Any air leaving the unit is filtered and cleaned before it leaves, so in environmental terms it is a very clean system. But it is also a very complex system, which makes the handling of the waste much more expensive.

No matter how it is made, it is the finished compost that provides the connection with urban agriculture, and indeed agriculture as a whole, for compost is really one of the most useful resources upon which agriculture can draw – especially organic production. Compost can be put to many uses, and distributed through a variety of market channels.

Most of the compost produced by centralised units, and of course all the compost produced in domestic units, is used by the domestic sector. This compost is suitable for use as a soil improver without any further treatment: it can be dug in before vegetables are sown or incorporated into the soil around the roots of newly planted trees. It can also be mixed according to balanced formulae with things like bark, or other wood fibre, to make a growing medium for container-grown plants: the unblended compost is rarely suitable for direct use as a potting compost. Compost can also be formulated with fibrous materials, such as bark, for use as a peat replacement.

If we consider the statistics on compost usage and likely supply, however, it is clear that once the targets are reached, very large volumes of compost will be generated that will very soon satisfy these domestic sector markets. Although every householder should be encouraged to use compost, we know that this market is not big enough to absorb all the compost that will become available. Indeed, it has been estimated that if all biodegradable waste were to be composted then the domestic sector market would be able to absorb only 2 per cent of the material produced. There must be further expansion, therefore, in the use of compost by the landscaping industry in urban parks and gardens. And, more importantly, some of this material will eventually go to agriculture. Agriculture is the secure, large-volume market for compost, and HDRA is working hard to promote its use on farms.

The horticultural benefits of compost can be summarised as follows. Compost is decomposed humic material with nutrients, and it is most widely used to improve the structure of soils. Heavy clay soil can be lightened with it, improving water infiltration and the crumb structure of the soil. In sandy soil compost binds the sand with other particles and increases water retention. All plant production – food production – is dependent on having a well-structured soil, and it is here, in the promotion of food production, that one of the important benefits of compost lies. Compost can also supply plant nutrients, both short-term and long-term, and therefore act as an effective replacement, or at least a partial replacement, for artificial fertilisers, and one that is more sustainable, given that artificial fertilisers very energy demanding.

But perhaps the biggest benefit of all lies in the microbiological activity which compost supports. Compost is full of microbes – a great diversity of organisms – and adding these microbes is good for the soil. But these microbes are in the compost in the first place because it contains energy, the sources upon which these organism rely to be active. Adding this energy source – the organic matter – to the soil therefore helps to increase the biological activity within the soil itself, with consequent improvements resulting to the structure of the soil, because active microbes secrete substances that help glue soil particles together. Furthermore, a microbially active soil can actually control some pests and diseases.

Nevertheless, some barriers remain to the greater use of compost in food production. As regards its use in agriculture, while this is potentially the largest market, there are problems related to the costs associated with using compost. Compost is a low-value product, and the main value associated with it is the saving in environmental costs derived by not sending its source material to landfill. But moving compost long distances away from the points of production will mean incurring new transport costs. We should therefore encourage the use of compost for food production within urban and peri-urban areas, as well as in the traditional urban markets detailed above, in order to keep these costs to a minimum.

Note: this chapter is a transcript of a lecture delivered by Dr Margi Lennartsson, Director of International Research at the Henry Doubleday Research Association. It has been edited by Richard Wiltshire.

PART THREE

PLANNING FOR CPULs:
OPEN URBAN SPACE

13

FOOD IN TIME: THE HISTORY OF ENGLISH OPEN URBAN SPACE AS A EUROPEAN EXAMPLE

Joe Howe, Katrin Bohn and André Viljoen

PLANNING FOR CPULs: OPEN URBAN SPACE

Figure 13.1

Figure 13.1 *1942 Albert Memorial, London.*

In Britain, as in much of the developed world, the very idea of growing food in the city, to many, sounds naive or even perverse. By contrast, urban food production in other parts of the world is a central feature of everyday life. For many poorer developing countries, urban agriculture is more a matter of economic value than of recreational or aesthetic preference (Lewcock, 1996). The scale of urban agriculture elsewhere is often staggering by Western standards. Across Chinese cities as a whole, 85 per cent of vegetables consumed by residents are produced within those cities and Shanghai and Beijing are fully self-sufficient in vegetables (Hough, 1995). This information may seem irrelevant to rich European countries. However, the degree to which attitudes towards urban food production are based on culture rather than wealth is demonstrated by the case of affluent Hong Kong. Here, vegetables to meet 45 per cent of local demand are grown on 5–6% of the total land area (Garnett, 1996).

In this chapter we will examine European cites to see why and when they have accommodated urban agriculture. We will do this by considering, in the main, the history of urban food growing in Britain. While this situation will not exactly match that found in other countries, it broadly follows a similar pattern and so the lessons learnt can be applied elsewhere.

WHERE WE LIVE IS WHERE WE GROW

Prior to the Industrial Revolution, the absence of sophisticated, high capacity transport systems and

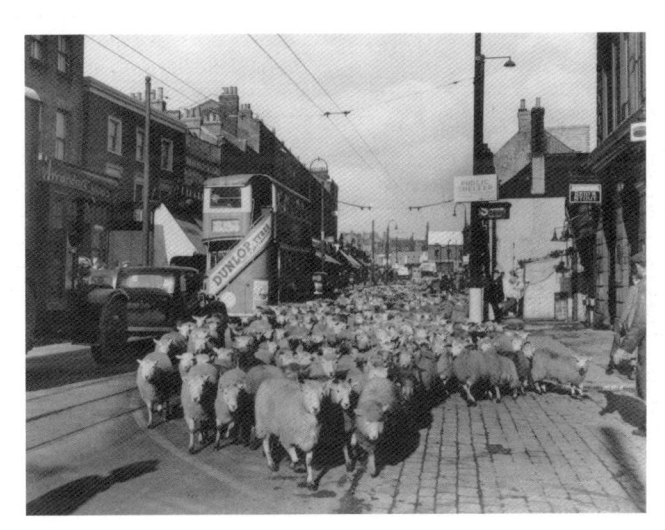

Figure 13.2

Figure 13.2 *Sheep being herded down a north London Street, November 1940.*

of preservation techniques such as refrigeration, inevitably meant that people had to grow food close to where they lived. Consequently, for thousands of years, built and cultivated environments co-existed: homes, markets, public buildings, and sacred places were interspersed with kitchen gardens, farms and common grazing land delivering food for the settlement's population. Towns and cities had distinct edges, often defined by city walls or by geographical features such as rivers or marshland. Open space devoted to food within the city's boundary might have been small and patchy, but urban settlements would supplement this by using a distinct rural area around their boundaries for food production.

Food that was grown outside built settlement boundaries presented few supply and transport problems given the small size of pre-Industrial Revolution settlements. Nearly all European towns and cities did not exceed 30 000 inhabitants living on an average 5 hectare urbanised area. Even London, as late as the seventeenth century, was contained almost entirely within a mile and a half radius, ensuring that virtually all of its inhabitants lived close to the countryside and to their source of food production.

INDUSTRIAL REVOLUTION AND SUBURBAN UTOPIAS: THE DIVORCE OF CITIES AND FOOD PRODUCTION

The close relationship between urban populations and food production largely fell apart during the Victorian Industrial Revolution. At first, despite dramatic population growth, poor transportation meant that the physical expansion of cities was limited. This changed from the mid-nineteenth century onwards with the building of the railways allowing people to live much further away from their work places. By the late nineteenth century, the sheer scale of the great industrial cities with their dense urban development and lack of green space separated millions of people from any immediate contact with food production.

The desperately unhealthy living conditions endured by urban factory workers and alienation from nature were causes of mounting concern. One important attempt to reintroduce open urban space was the development of the municipal park, adopted for example in the UK, in the second half of the nineteenth century (Nicholson-Lord, 1987) (see Chapter 14). The other was the spread of urban allotments during the same time.

Most large parks in European cities date from the nineteenth century. Their often picturesque landscapes, modelled on the open countryside, forests or

Figure 13.3

Figure 13.3 *A wartime roof garden in London.*

feudal property, attracted the city dweller for leisure activities. Leisure was free, with outdoor pursuits like walking, sitting, reading, picnicking, ball games, ice skating, etc., all with a choice of minor commercial involvement in the form of small entrance fees or ice cream. The park's significance as leisure space has remained unchanged since and is still associated with similar activities, though leisure activity now takes more diverse forms (see Part Five).

Smaller urban settlements often faced fewer spatial problems during industrialisation and, in consequence, did not develop inner-city urban parks of significant scale. In these towns and cities it was the urban periphery that provided open urban space for city dwellers, both in the form of open park landscape (often agricultural/farmland) adjacent to the settlement and house gardens and allotments in the developing suburbia.

Allotments were originally established in the early eighteenth century to compensate the landless rural poor for the enclosure of common land by wealthy landowners (Crouch and Ward, 1988). Their function was to provide a nutritional and economic safety net against unemployment or to supplement meagre incomes. The need for urban allotments arose soon after this and became acute during the nineteenth century as the trickle of landless poor migrating to the great cities became a flood. Allotment provision at this point was largely private and ad hoc.

But by the late nineteenth century the growing power and responsibilities of local government were reflected in the first allotments legislation. Acts of 1887 and 1892 required local authorities for the first time to provide allotments for the labouring poor where need was shown. These Acts were consolidated and the requirement for municipal provision strengthened in the 1908 Small Holdings and Allotments Act. This Act remains the principal piece of legislation governing allotments (Crouch and Ward, 1988). Comparable developments were occurring in other parts of Europe, for example, the introduction of Schrebergärten in Germany.

Urban food growing in general, and allotments in particular, featured prominently in Ebenezer Howard's *Garden Cities of Tomorrow* first published in 1898. Howard's ideas drew on an earlier 'back to the land tradition' embodied in examples such as Bournville model village built by enlightened industrialist George Cadbury for his workers near Birmingham in 1895 (Marsh, 1982).

Howard's garden city envisaged the planned dispersal of the population from the overcrowded slums of the great industrial cities of Britain to new towns. These were to be located beyond a green belt, separating them and their 'parent' city. Each was to have a population of approximately 30 000 and be grouped around a larger central city in polynuclear fashion – the whole creating a 'social city.'

Food production within or around Howard's garden cities was a key element. In each city, five-sixths of the area was devoted to food production. Residential space was to be divided into generous plots of 20 by 130 feet, which Howard envisaged would be sufficient to feed a family of at least five people. In addition, allotments ringed the settlement peripheries. Twentieth century garden city derivatives like Welwyn and Letchworth never quite developed into the self-sufficient, food-producing entities originally conceived. Notwithstanding this, 33 new towns were built (Ward, 1993), with at least an ambition to integrate landscape and living space.

While Howard's theories had far reaching effects on town planning across Europe, it is probably true to say that the town planning theories of Le Corbusier, expounded in '*The City of Tomorrow and its Planning*' (Le Corbusier, 1971) and first published in 1924, had the greatest international influence on the architecture and urban planning of the twentieth century.

Le Corbusier's attitude to Howard's theories are well summarised by Maurice Besset:

> Le Corbusier's debt to the theorists of the garden city is equally certain and no less ambiguous. Of course, he spoke out against the danger of 'de-urbanisation' which the garden cities represented, for they proposed a false solution to the city and could only lead to 'a sterile isolation of the individual,' whom they would maintain in 'a slavery organised by capitalist society.' As he fought against the 'corridor-street' and the slums it gives rise to, so he fought against the 'great illusion' of the individual home, taking up space and generating circulation. To the horizontal garden city he opposed as early as 1922 his large, regenerated apartment building, his vertical garden city. This apartment block, consisting in essence of superimposed villas, this immeuble-villa as he called it, reduced distances and facilitated social contacts and integration of the different urban functions. But whether the dwellings were grouped vertically or horizontally, it was still a garden city and the same essential joys promised by Howard and his friends towards which sails, above the trees of the green belt, the great concrete ship of the apartment block which he called the Unité de Habitation, the housing unit.
>
> (Besset, 1987)

Although not named as such, Urban and Peri-Urban Agriculture play a central role in Le Corbusier's urban thinking. In Chapter 13 of 'The City of Tomorrow', under the heading 'Concerning Garden Cities' he describes precisely how urban agriculture could be accommodated without reducing the overall density of suburbs. Analysing a typical suburban housing plot of 400 m^2 he proposes allocating 150 m^2 to a communal market garden. 'There would be a farmer in charge of every 100 such plots and intensive cultivation would be employed . . . Orchards lie between the houses and the cultivated land' (Le Corbusier, 1971).

Later, Le Corbusier wrote about what would today be called Peri-Urban Agriculture. By 1945, he had defined 'three human establishments': The Farm Unit, The Linear Industrial City, and the Radio-Concentric Change-Over City. What we find interesting today, about these proposals, is the way they present a series of overlaid networks. Within this conception of the city, agricultural land provides a kind of underlying carpet across which linear cities between 50 and 200 kilometres long form strands in a network, the nodes of which become radio-concentric change-over cities. The clear boundaries between farm units and cities are symptomatic of a prevailing interest in zoning (Besset, 1987). If we review these proposals today, Le Corbusier's Farm Unit would be considered peri-urban agriculture. The triangulated network of linear cities he sketched would result in food growing for cities within the limits currently set by the manages of London's farmers' markets (see Chapter 10).

Taking a different position to Le Corbusier's, the North American architect Frank Lloyd Wright published during the middle of the twentieth century and towards the end of his life a series of essays, eventually brought together in his book, '*The Living City*.' Wright's vision of the living city could best be summarised as integrating agriculture into dispersed suburban settlements. The living city may be read as a tirade against what Wright saw as dehumanising aspects of 'purism,' the architecture emerging from Europe in response to the machine age. Much of the tone of the volume appears to despair at the consequences of conflict and a non-human-centred economy: '. . . instead of practising democracy, we now defend only what we call our interests. So we go from war to war.' (Lloyd Wright, 1970).

In common with Le Corbusier's vision, Wright's living city celebrates personal transport, although in a somewhat eccentric manner. But within Lloyd Wright's proposition, we find a vision that resonates with current architectural thinking about the essence and generative power of the concept of landscape: 'Architecture and acreage (agricultural land) will be seen together as landscape, as was the best in antique architecture, and will become more essential to each other.' (Lloyd Wright, 1970).

This notion of 'Architecture and acreage, seen together as landscape,' is perhaps Wright's greatest gift to contemporary architects and urbanists. As an idea, it frees us from distinctions between urban and suburban, and also helps to articulate a vision of a city driven by Ecological Intensification, where productive landscapes can stand equal to traditional development in the built environment. While architects have started to deal with landscape and building, informing each other at an architectural scale, variously referred to as earth buildings, groundscrapers, landscrapers or subscrapers (Brayer and Simonot, 2003; Betsky, 2002; Richards, 2002) at the scale of a city, these issues are only beginning to be addressed.

URBAN FOOD AND CONFLICT

But it was not the ideas of architects that had the most dramatic effect on urban agriculture in Britain and Europe between 1900 and 1945 – the biggest stimulant to urban food production in Europe was undoubtedly war.

In both World Wars, the real threat of starvation posed by blockades prompted campaigns to increase indigenous food output, much of it from urban agriculture. In the First World War, serious UK government campaigning to boost food production did not start until 1917 for fear, as with rationing, of damaging civilian morale. Despite this, campaign results were dramatic. The number of allotments, each typically 250 m^2 in area, roughly tripled from between 450 000 to 600 000 in 1913 to between 1 300 000 and 1 500 000 by late 1917, which together produced 2 000 000 tons of vegetables (Crouch and Ward, 1988).

Between the wars, interest in allotments and other forms of urban food growing declined throughout Europe, although it never fell back to pre-1914 levels. Mass unemployment from the late 1920s onwards sparked a revival in allotment interest as a valuable means of self-help. Philanthropic groups of various kinds, in Britain notably the Society of Friends, ran schemes providing fertiliser, seeds and tools. Currently, similar support mechanisms have been introduced in Cuba, to support its national programme of urban agriculture (see Chapter 17).

At the outbreak of the Second World War, the UK government was determined not to repeat the mistake of the previous conflict in leaving preparations for boosting home food production too late. Accordingly, the famous 'Dig for Victory' campaign was launched by the Minister of Agriculture in October 1939. As in 1917–18, the results of the campaign were impressive. By the middle of the war, a survey showed over half of all manual workers were producing food from either an allotment plot or garden, and by the war's end there were approximately 1 500 000 allotments. In 1944 these, together with gardens and other plots of land, including parks turned into fields, were meeting fully ten per cent of national food needs and around half the nation's fruit and vegetable requirements (Crouch and Ward, 1988). In addition to fruit and vegetables, livestock clubs also sprang up in abundance.

PLANNING FOR CPULS: OPEN URBAN SPACE

Figure 13.4

Figure 13.4 *1939–1945, The Tower of London.*

Figure 13.5 *1939–1945, Sunday Morning, Clapham Common, London.*

URBAN REBUILDING AND URBAN FOOD DECLINE

In Britain, the end of the 'Dig for Victory' campaign was followed by a sharp decline in urban food production. Throughout the 1950s and 1960s, a great deal of food growing land was returned to its original pre-war use or lost to new development. The combined effect of the new welfare state, effectively full employment and increasing prosperity meant that people no longer saw a need to grow their own food. Furthermore, allotments suffered from an image problem: they were associated with wartime austerity and 'make do' and certainly did not chime with the spirit of an age dedicated to scientific progress and youth culture.

To address the issues surrounding the decline in allotment use, the incoming Wilson Government set up an inquiry chaired by Professor Harry Thorpe in

Figure 13.6

Figure 13.7

Figure 13.6 *Children digging in their school's vegetable garden.*
Figure 13.7 *Girton college students during the Second World War.*

1964. Thorpe appreciated the value of allotments, but was deeply critical of the allotment movement as a whole. He believed that the best hope of allotment survival in the affluent new age lay in re-creating them as 'leisure-gardens,' which avoided the charitable associations of the term 'allotment.' Thorpe also observed that the legislation surrounding allotments was highly confusing. Amongst his 44 recommendations, he called for a new allotment Act to rationalise the law. However, to date, none of Thorpe's recommendations have been taken up by government. Thorpe's proposals are by no means universally admired (see Chapter 15), critics arguing that one of the consequences of adopting them would have been to deny the informal and self-directed use of allotments, which is essential to their character and function in the city. But to deny the positive aspects of Thorpe's recommendations is equally short-sighted. Indeed, the observations made by the authors, of layers of occupation at the Moulsecoombe allotment near Brighton (see Chapter 19) show how allotments are already being used for leisure. If we think of allotments as one component of productive urban landscape, set within continuous landscape, then Thorpe's leisure concepts may be seen as prompts for a new understanding of leisure landscapes.

REVIVAL AND DIVERSIFICATION OF URBAN FOOD GROWING

Since the 1970s, environmental awareness groups backed by local authorities, public developers and community enthusiasts have also been battling to safeguard, promote or improve open urban space. Whilst this might have been less important for the established urban parkscape, it encouraged a different approach to the design of new types of open space, often benefiting smaller, underused or formerly industrial (brownfield) inner-urban sites.

The early 1970s marked a change in fortune for allotments and brought new forms of urban food growing activity in Britain. The main reason seems to have been a growing environmental ethic, developed initially during the 1960s, as alternative lifestyles, and notions of self-sufficiency supported by the use of alternative (renewable) energy led to a renewed appreciation of urban food production. The effect of this culture shift was to greatly reduce the rate of allotment loss in Britain (down by 84 per cent from 1970 to 1977) and sharply increase the demand for allotments in many places. This burgeoning environmental awareness also gave rise to the development of new forms of urban food growing activity – notably the urban farm and community garden movements.

In Britain, the first urban farm was started in Kentish Town, North London, in 1971, and by the 1990s there existed more than 60 such farms all over the country (Hough, 1995). Urban farms are about more than simply bringing the countryside into the city. They are invariably located in poorer districts and the most successful have tended to act as a focus for environmentally conscious urban regeneration. Within this broad framework, most farms are multi-purpose entities, although environmental education is always a strong theme. Farms generally keep livestock for food, for educational purposes and for their other products, as well as running garden or allotment plots. Many have a commercial aspect with craft workshops, shops and restaurants selling farm produce and offering activities such as horse riding. They also provide venues for public meetings and frequently run training courses.

The closely related concept of community gardens originated in the USA in the early 1970s. Like urban farms, what characterises community gardens and distinguishes them from traditional allotments, is the emphasis on group activity and their role as a focus for community regeneration. They tend to differ from urban farms in being generally smaller and in not having livestock, although exceptions exist. The movement has often been strongest in deprived urban districts, in places like The Bronx and Harlem in New York, where women, particularly black women, are especially prominent as activists and participants (Hynes, 1996). In 1978, the American Community Gardening Association (ACGA) was formed and the movement has since grown dramatically. Between 1990 and 1992, the ACGA reported the setting up of 523 new community gardens in 24 cities across the USA (Hynes, 1996). In New York alone, the Green Thumb Community Gardening Programme has developed from its inception in 1978 to encompass 700 community groups across the city by the mid-1990s (Garnett, 1996). The community gardens concept has now become established in Britain, and is represented by the Association of City Farms and Community Gardens (see Chapter 11).

URBAN AGRICULTURE AND SUSTAINABILITY

The emergence of the idea of sustainability, which was a defining feature of the 1992 Rio Earth Summit, was instrumental in raising environmental awareness and provided a powerful rationale for reassessing contemporary design and development strategies.

Within architecture, the major impact was on finding ways of reducing the energy consumption of buildings and thereby reducing their greenhouse gas emissions. As a result of investment in marketing, publications, competitions and demonstration projects, by, for example, the European Union, Professional Bodies, National Institutions and Investors, most architects have become aware of the factors contributing to sustainable building design. At a national level, building codes have been amended to improve the energy efficiency of new development.

By contrast, the environmental benefits of sustainable landscape design have received less extensive publicity, although by now concepts such as ecological corridors are well established. Publications in this field are becoming more common as the importance of an integrated view of urban development is appreciated (Santamouris, 2001; Thompson and Sorvig, 2000).

The significance of urban agriculture within contemporary open urban space is highly variable according to the city examined. The environmental benefits of urban agriculture are only now beginning to be identified and acknowledged, and currently its significance is very different in developed and developing countries. In developing countries, urban agriculture is largely driven by ecomic need, while in developed countries it is more likely to have arisen in response to social or recreational needs and desires (see Part Four – Planning for CPULs: International Experience). In Europe, interest in allotment holding, urban farming or community gardening has constantly increased in recent years, with a resulting resurgence of urban food growing.

Taking Britain as an example, it is true to say that there has been nothing like the investment in urban landscapes which took place in the nineteenth century with the development of large municipal parks (see Chapter 13). In Britain, there has been little celebration of contemporary urban landscape until very recently. Some new interventions, i.e. those in the east of London (Mile End Park and Thames Barrier Park, see Chapter 14 – Open Urban Space Atlas) have now received positive publicity. On the other hand, significant work is currently being carried out within Milton Keynes, one of the new towns which was built during the twentieth century as a consequence of Howard's garden city theories.

Milton Keynes retains and is developing a legacy of public parkland, using innovative management, in support of the public interest. This supports one of Howard's ambitions, which was to avoid inflated land prices and thus retain open urban space. His notion was that new town companies would purchase land for development, rent houses to residents and, once the land was paid for, invest rents in the improvement of the public realm. Within Milton Keynes farsighted plans integrating parkland into the town, and a contemporary financial and management strategy are supporting open urban space and gaining recognition:

> ... The single statistic that it has planted twenty million trees, and the fact that it has inside its boundaries the largest, most diverse park system of any city in the UK, should be enough to change perceptions... The parks are not the monotonous, scrubby, flat playing fields with moth eaten flowerbeds that are all many underfunded municipalities seem to be able to afford.... This is because Milton Keynes's parks are run not by the cash-strapped council, but by a charitable trust. The trust is endowed with property assets worth 50 million pounds, originally including 14 city pubs and an industrial estate from which it derived an income of 3.1 million pounds last year to fund its activities.
>
> (Brown, 2003)

The dedicated non-profit-making charity established by Milton Keynes in 1992 to manage its large landscape interventions has fared better than the council in maintaining open urban space (Brown, 2003). Milton Keynes provides a financial model for the support of CPULs.

REFERENCES

Besset, M. (1987). *Le Corbusier: to live with light.* Architectural Press.

Betsky, A. (2002). *Landscrapers: Building with the land.* Thames and Hudson.

Brayer, M. and Simonot, B. (eds). (2003). *Archilabs earth buildings: radical experiments in land architecture.* Thames and Hudson.

Brown, P. (2003). Parkland. *The Guardian: Guardian Society*, 23 April, p. 8.

Crouch, D. and Ward, C. (1988). *The Allotment.* Faber and Faber, London.

Garnett, T. (1996). Harvesting the cities. *Town and Country Planning*, **65**(9), 264–265.

Hough, M. (1995). *Cities and natural process.* Routledge, London.

Hynes, P. (1996). *A pinch of eden.* Chelsea Green, White River Junction.

Le Corbusier. (1971). *The city of tomorrow and its planning, 3rd Edn.* Architectural Press.

Lewcock, C. (1996). Agricultural issues for developing country city management. *Town and Country Planning*, **65**(9), 267–268.

Lloyd Wright, F. (1970). *The living city.* Meridian Books.

Marsh, J. (1982). *Back to the land.* Quartet, London.

Nicholson-Lord, D. (1987). *The Greening of Cities.* Routledge, London.

Richards, I. (2002). *Groundscrapers + subscrapers.* John Wiley & Son.

Santamouris, A. B. (2001). On the built environment – the urban influence. In *Energy and Climate in the Built Environment* (A. Santamouris, ed.) pp. 3–18, James and James.

Thompson, J. and Sorvig, K. (2000). *Sustainable Landscape Construction.* Island Press, Washington DC.

Ward, C. (1993). *New Town, New Home, The lessons of experience.* Calouste Gulbenkian Foundation.

14

FOOD IN SPACE:
CPULs AMONGST CONTEMPORARY OPEN URBAN SPACE

Katrin Bohn and André Viljoen

CPULs FOR EUROPEAN CITIES

In most European cities, movements towards the revitalisation of city centres have pushed the importance of open urban space to the front of public awareness. The resulting open urban spaces satisfy an immense variety of desires and programmes. Their themes are as different as their clients, occupants or locations, culminating in a multitude of approaches and solutions. Cities and towns are now full of fantastic new and old urban squares, urban parks, urban riverfronts, urban stages, urban forests and urban beaches. *There are no urban fields yet.*

CPULs will *embed* fields, thereby enriching urban tissue and lifestyle and contributing actively to solving environmental problems. Whether differing from or resembling, these productive landscapes will exist alongside other open urban space designs (see Open Urban Space Atlas, Figures 14.1–14.11). CPULs as urban design strategy could act as moderator between local user desires and strategic urban planning, between social and economic viability, sustainable ideas and urban productivity, between short-term advantage and long-term benefit.

CPULs could formally be very similar to *urban parks* in that both are mainly natural and designed to certain spatial or functional criteria. They will also be similar to *urban forests*, which again are natural, though of physically denser or higher vegetation. Urban parks and forests will be agriculturally less productive, but can therefore allow a freer use of space as there will be no movement restrictions which need to be designed in CPULs to protect crops. In that respect, CPULs could be similar to *urban gardens* as both will follow particular planting sequences and patterns. CPULs will mostly be larger than urban gardens and not as contained by user-specific or ornamental design.

Urban squares will be most dissimilar to CPULs. Compared to their free use as social and event spaces, CPULs will need to guide people and events to allow for undisturbed working and growing on the fields.

Urban river fronts (urban beaches) are seen as open urban spaces along a natural boundary. Being also about movement, they are however, in layout and function more similar to urban parks or squares than to CPULs. Urban beaches, as well as urban squares or parks, can turn into *urban stages*, flexible-use open spaces reflecting new attitudes to leisure. As those might be of any layout, they could become temporarily similar to CPULs.

Urban routes facilitate movement in the city and can also be of any form, layout, size and often of ingenious construction. They will resemble CPULs in their emphasis on movement.

THE OPEN URBAN SPACE ATLAS

For comparison with other types of open urban space, we drew on three criteria which embrace, in our eyes, the most important qualities of a CPUL:

Spaciousness – as its inheritance, Occupation – as its present, and Ecology – as its gift to the future.

Spaciousness

Spaciousness describes the space itself, its extent, its width and breath. It means more than *size*, but size is its basic element, its starting point. There is no qualitative judgement connected to size: small open space is not bad open space, neither is big open space. Size is considered as influencing the space's designation and its ability to accommodate certain programmes and occupants. Size is very manipulable by design, involving topography, view axis, walk axis, change in vegetation, building, etc. *Sense of openness*, though connected to size, reflects this manipulation by providing a more sensual, qualitative measure for the spatial success of open urban space. It relates to the occupation and function of the space as well as to its position in the urban grid and its connectivity. Integration into inter-connecting urban routes enhances the significance of individual open space within any urban network. To be able to walk continuously onwards from an open urban space extends the space beyond itself and into a very fine and slow layer of *inner-city movement*. The potential for such movement encourages occupation and occupants as well as shaping the form and layout of open urban space. It also introduces change and renewal to the space, therewith offering a particular *persistent visual stimulation*. Stimuli can be drawn from a variety of sources, such as occupancy (events, activities, movement, etc.), but they are mostly strongly linked to the natural environment: to vegetation undergoing seasonal change, growth, changing planting pattern, to water, wind, sun, rain, etc. Visual stimulation refers back to spaciousness, as a visually stimulating space is more likely to be judged as appropriately sized.

Occupation

Occupation is one of the prime concerns when planning contemporary open urban space. 'Occupation' thereby is often rather tightly measuring the success of new designs by quantifiable criteria such as the number of people flocking in during event time or sunshine (i.e. communal parks, fun parks, theme parks including traditional theme parks, such as zoos), or by the financial turnover of on-site facilities (i.e. leisure centres, shops, eateries, eating for leisure, leisure to reinstate health, beauty or body shape).

A more holistic view of current occupation of open space could include qualitative and longer-lasting criteria, such as education, health, potential for communication/integration or personal enrichment, i.e. the satisfaction about own individual action and its importance for a broader urban community. Open urban spaces focusing that way will develop strong *local interactions* enabling the space to attract and retain local people.

Depending on its programme and tolerance to change, an open space providing for local interactions will very likely accommodate a large *variety of occupants*. If asked, occupants appear less interested in a space's size or location and more in its potential to let them integrate and participate (see Chapter 15). We judge attracting large varieties of occupants to be as important as attracting large numbers of them. The latter suggests specific sizes/layouts of open urban space as well as specific programmes to enable the entertainment of many occupants. It is useful to look at the *variety of occupation* on offer, not only to measure the success of open urban space to continuously attract people, but also to establish how the logistics of large user numbers are managed and whether any income can be generated from that.

Apart from income generation through on-site facilities, the *economic return from ground-use* is considered a major factor in judging the long-lasting success of open urban space. As its ground is the very starting point of any open urban space and as this ground is rare, it becomes important to measure how it is treated.

Ecology

In our context, 'ecology' gives weight to open urban space by connecting present design and programme to a widely desired and sustainable future. It also proposes a strategy to manage that process.

Ecology in open urban space is great where peoples' exposure to *urban nature* is great. Commonly,

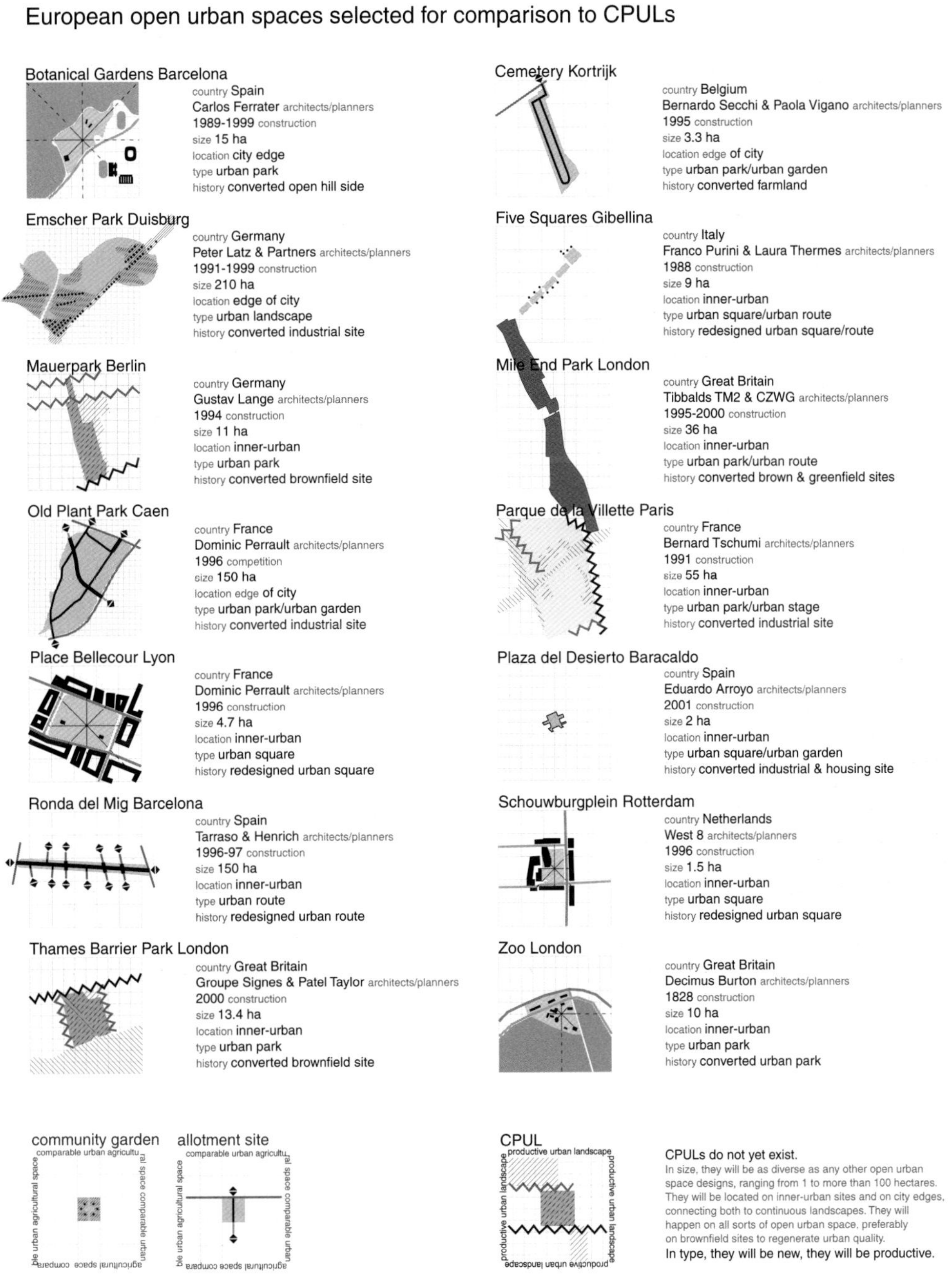

FOOD IN SPACE

European open urban spaces selected for comparison to CPULs

Botanical Gardens Barcelona
country **Spain**
Carlos Ferrater architects/planners
1989-1999 construction
size **15 ha**
location **city edge**
type **urban park**
history **converted open hill side**

Emscher Park Duisburg
country **Germany**
Peter Latz & Partners architects/planners
1991-1999 construction
size **210 ha**
location **edge of city**
type **urban landscape**
history **converted industrial site**

Mauerpark Berlin
country **Germany**
Gustav Lange architects/planners
1994 construction
size **11 ha**
location **inner-urban**
type **urban park**
history **converted brownfield site**

Old Plant Park Caen
country **France**
Dominic Perrault architects/planners
1996 competition
size **150 ha**
location **edge of city**
type **urban park/urban garden**
history **converted industrial site**

Place Bellecour Lyon
country **France**
Dominic Perrault architects/planners
1996 construction
size **4.7 ha**
location **inner-urban**
type **urban square**
history **redesigned urban square**

Ronda del Mig Barcelona
country **Spain**
Tarraso & Henrich architects/planners
1996-97 construction
size **150 ha**
location **inner-urban**
type **urban route**
history **redesigned urban route**

Thames Barrier Park London
country **Great Britain**
Groupe Signes & Patel Taylor architects/planners
2000 construction
size **13.4 ha**
location **inner-urban**
type **urban park**
history **converted brownfield site**

Cemetery Kortrijk
country **Belgium**
Bernardo Secchi & Paola Vigano architects/planners
1995 construction
size **3.3 ha**
location **edge of city**
type **urban park/urban garden**
history **converted farmland**

Five Squares Gibellina
country **Italy**
Franco Purini & Laura Thermes architects/planners
1988 construction
size **9 ha**
location **inner-urban**
type **urban square/urban route**
history **redesigned urban square/route**

Mile End Park London
country **Great Britain**
Tibbalds TM2 & CZWG architects/planners
1995-2000 construction
size **36 ha**
location **inner-urban**
type **urban park/urban route**
history **converted brown & greenfield sites**

Parque de la Villette Paris
country **France**
Bernard Tschumi architects/planners
1991 construction
size **55 ha**
location **inner-urban**
type **urban park/urban stage**
history **converted industrial site**

Plaza del Desierto Baracaldo
country **Spain**
Eduardo Arroyo architects/planners
2001 construction
size **2 ha**
location **inner-urban**
type **urban square/urban garden**
history **converted industrial & housing site**

Schouwburgplein Rotterdam
country **Netherlands**
West 8 architects/planners
1996 construction
size **1.5 ha**
location **inner-urban**
type **urban square**
history **redesigned urban square**

Zoo London
country **Great Britain**
Decimus Burton architects/planners
1828 construction
size **10 ha**
location **inner-urban**
type **urban park**
history **converted urban park**

community garden
allotment site
CPUL

CPULs do not yet exist.
In size, they will be as diverse as any other open urban space designs, ranging from 1 to more than 100 hectares. They will be located on inner-urban sites and on city edges, connecting both to continuous landscapes. They will happen on all sorts of open urban space, preferably on brownfield sites to regenerate urban quality.
In type, they will be new, they will be productive.

Figure 14.1

PLANNING FOR CPULs: OPEN URBAN SPACE

Comparative size study . CPUL near Victoria Park, London, *Bohn & Viljoen 2001*, see --> Practical Visioning

legend
□ urban square
■ urban garden
■ urban park

scales
□ 100m x 100m = 1ha
10ha x 10ha = 100ha

size

Most qualities of open urban space which occupants value are not quantifiable. Size is, but in an ambiguous way: to look at size in quantifiable terms is at the same time necessary and completely superfluous. The diagrams depict absolute size, which is crucial, f.e., for comparison of the site's dominance in the urban grid, esp. when looking at open space that superficially seems to be similar *(ie CPUL-urban agriculture, allotment site)* or seems to have potential site yields (see financial return from ground-use) *(ie CPUL-urban agriculture, Zoo)*.

The difference between the two possibly most popular open urban leisure spaces in Europe - the urban square and the urban park - is not only vegetation or ground cover but size. Dissimilar typologies lead here to diverse user expectations of size: squares are imagined and accepted as much smaller then parks *(ie Schouwburgplein, Mauerpark)*.

Through design, the sensed size of open urban space can become very different from its physical size: topography and proportion can visually enlarge open space, high and/or dense boundaries reduce it *(ie Mile End Park, Plaza del Desierto, Schouwburgplein)*. The size of the surrounding city also manipulates the open urban space size, in that big cities absorb their open spaces whereas small cities expose them *(ie Mauerpark, Caen)*.

local interactions

Memory is what keeps space alive. Space that creates memories is space of encounters. Whilst these do not necessarily need to be encounters between people, they mostly are. Spaces allowing encounters will both attract occupants and keep them coming back.

Every designed open urban space is designated to a certain theme. If the designation incurs separation from parts of the public, the space will allow interactions only between very specific user groups *(ie Zoo, allotment site)*. Even if not separated, open space can be confined to specific user groups only, if located in particular areas or designed in particular ways *(ie Old Plant Park)*. Whilst these spaces often work well in themselves, they do not reflect the idea of *open* urban space.

As soon as open urban space allows a certain indetermination, it will be used by varieties of people, engaging in varieties of interactions. When usable as meeting place, the space becomes a source of even more diverse actions, as people are setting out from it and into it. Most of these interactions are independent of the space's size, layout or finish *(ie Schouwburgplein, Mauerpark, Plaza del Desierto)*. New quality will be reached when interactions between working, trading and shopping people will happen alongside the leisure and service activities *(ie CPUL, Mile End Park)*.

Figure 14.2

FOOD IN SPACE

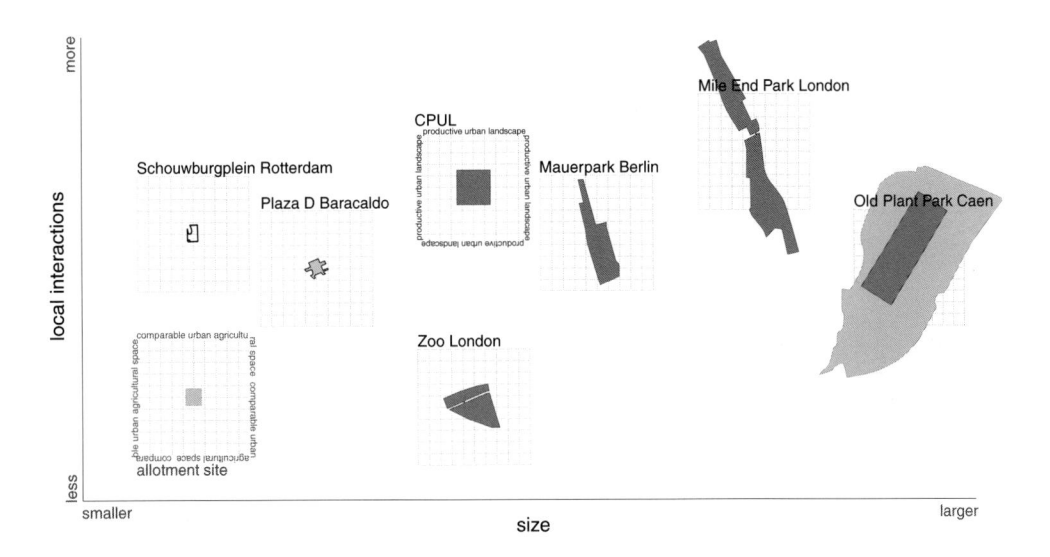

local interactions

Zoo London
one-off encounters between tourist visitors and/or local visitors and/or workers/traders

allotment site
repeated encounters between members of 2 specific user group (allotment holders, passers-by/through)

Old Plant Park Caen
repeated encounters between members of several specific user groups (residents, visitors, passers-by)

Plaza del Desierto Baracaldo
repeated and one-off encounters between members of various user groups (residents, tourist/local visitors, passers-by/through). meeting place

Schouwburgplein Rotterdam
repeated and one-off encounters between members of various user groups (residents, tourist/local visitors, passers-by, traders). meeting place

Mauerpark Berlin
repeated and one-off encounters between members of various user groups. meeting place

CPUL
repeated encounters between members of various user groups (residents, tourist/local visitors, passers-by) and/or workers/traders. meeting place

Mile End Park London
repeated and one-off encounters between members of various user groups (residents, tourist/local visitors, passers-by) and/or traders. meeting place

size

Schouwburgplein Rotterdam
1.5 ha. regular-size, flat
urban square in medium-city context

allotment site
1.5 ha. regular-size, flat or undulating
urban garden in any city context

Plaza del Desierto Baracaldo
2 ha. regular-size, stepped
urban garden/urban square in small-city context

Zoo London
10 ha. large-size, flat
urban park in big-city context

CPUL
10 ha. large-size, flat or undulating
urban park in big-city context. can also be much smaller, i.e. 1.5 ha, Schouwburgplein-size

Mauerpark Berlin
11 ha. large-size, undulating
urban park in big-city context

Mile End Park London
36 ha. very large-size, flat
urban park/urban route in big-city context

Old Plant Park Caen
150 ha. enormous-size, flat
urban park/garden holding incremental out-of-town housing

Figure 14.3

PLANNING FOR CPULs: OPEN URBAN SPACE

legend

■ solid edge
ie building

▪ see-through edge
ie fence, vegetation,
main traffic, water

unobstructed /
semi-obstructed
openness

scales   □ 100m x 100m = 1ha

□ 100m x 100m = 1ha

□ 100m x 100m = 1ha

Increasing openness . CPUL near East Street Market, London, *Bohn & Viljoen 2002*, see --> Practical Visioning

sense of openness

Sense of openness is desirable. It is one of the qualities people look for when travelling to the sea, climbing mountains or towers, cherishing a garden or balcony. Sense of openness means both, physical openness to the wandering/moving through space and sensed openness to people's wandering eyes (view) and ears (sound).

The *physical openness* of an urban space is determined by its physical boundaries: buildings, fencing, roads, canals, dense vegetation etc. How users *sense openness* depends on their personal experiences as well as on the design of the space: boundaries, ground, topography, internal objects, axes... manipulating the space's absolute size to appear bigger or smaller than it physically is. Buildings, f.e., often appear as the most restricting boundaries, both to eye and movement *(ie Place Bellecour, community garden, Schouwburgplein, Zoo)*. The sheer size or topography of open urban space can outweigh a boundary, ie an adjacent motorway, which for a small open urban space would pose a major border to both, movement and senses *(ie Plaza del Desierto, Emscher Park)*. Sensed openness also includes cultural references: two identical spaces, one enclosed by metal fencing, the other by dense fruit bushes will attract different occupants and/or usage *(ie Botanical Gardens, Place Bellecour, CPUL)*.

variety of occupation

Diversity of occupation, of activities is a measure for the success of open urban space in that it will create and maintain local interest, identity and satisfaction.

Occupation of open urban space can be everything from fast and short, ie moving through *(ie Place Bellecour)*, to slow and extended, or mental, ie occupying territory by sitting there and/or listening to / looking at it *(ie Emscher Park, Botanical Gardens)*. Spaces which attract a wide range of users mostly register a large variety of occupation *(ie Emscher Park)*. Occupation through a typical user / user group can happen once *(ie Zoo)* or repeatedly *(ie community garden)*. Indetermined design and/or the availability of open-minded space will encourage users to engage in unforeseen individual activities *(ie Schouwburgplein, Plaza del Desierto)*.

Contemporary open urban space invites mainly leisure *(ie Plaza del Desierto, community garden)*, using-service *(ie Schouwburgplein)* or educative activities *(ie Botanical Gardens)*. All of those activities can also be related to particular historic urban spaces *(ie Zoo, Place Bellecour)*. Open urban space achieves a non-precedented occupational quality when including working activity, whereby "working" means working *with* the actual space, ie earning a living by using elements inherent to the space: ground, vegetation, buildings *(ie CPUL-urban agriculture, Emscher Park)*.

Figure 14.4

114

FOOD IN SPACE

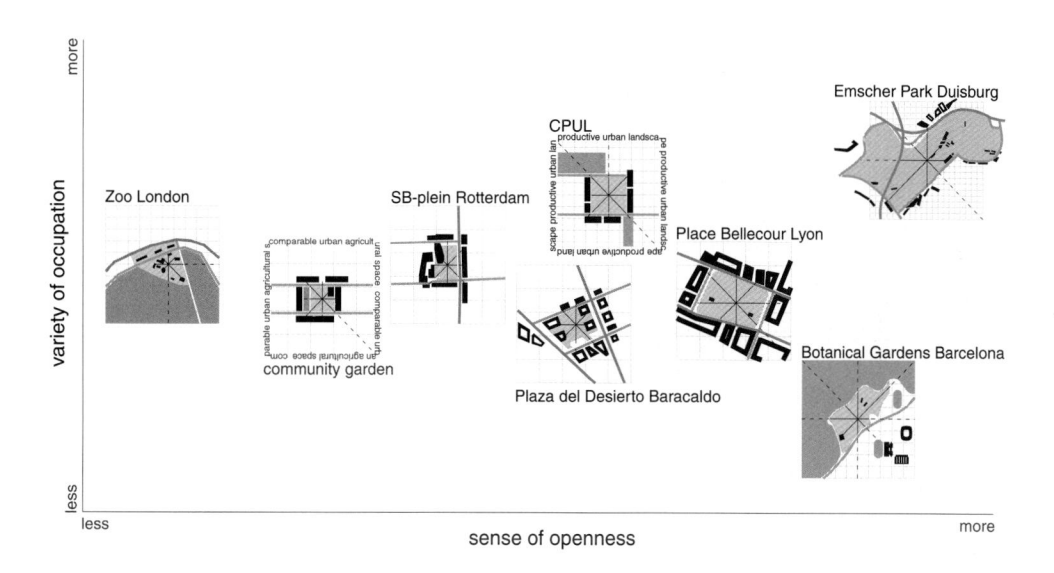

[Scatter plot with axes "variety of occupation" (less to more, vertical) and "sense of openness" (less to more, horizontal), showing positions of: Zoo London, SB-plein Rotterdam, community garden, CPUL, Plaza del Desierto Baracaldo, Place Bellecour Lyon, Botanical Gardens Barcelona, Emscher Park Duisburg]

variety of occupation

Botanical Gardens Barcelona
walk, talk, see, sit, observe, listen, learn
Zoo London
walk, talk, see, sit, observe, listen, learn
see shows/exhibitions, eat, play
Plaza del Desierto Baracaldo
walk, talk, see, sit, eat, play
short cut, meet, party
Place Bellecour Lyon
walk, short cut, talk, see, sit, meet, eat, play, party
cycle
community garden
walk, talk, see, sit, meet, eat, play, party
engage, work
Schouwburgplein Rotterdam
walk, cycle, short cut, talk, see, sit, learn
see shows/exhibitions, meet, eat, play, party
CPUL
walk, cycle, short cut, talk, see, sit, observe, learn
meet, eat, play, party, engage, work
produce, earn
Emscher Park Duisburg
walk, cycle, short cut, talk, see, sit, observe, listen
learn, see shows/exhibitions, meet, eat, play, party
engage, work

listed are groups of activities, i.e. there are various activities that can be done sitting

sense of openness

Zoo London
reducing sense of openness: internal buildings, internal fencing, fenced border, external roads, external dense trees
increasing sense of openness: large size
community garden
reducing: fenced border, external buildings and roads
increasing: open access
Schouwburgplein Rotterdam
reducing: external buildings and roads, road noise
increasing: no fencing/border, flatness, sky views
Plaza del Desierto Baracaldo
reducing: external buildings, roads and train
increasing: topography, one side open to views
CPUL
reducing: external buildings and traffic
increasing: topography, two sides open to views and moves
Place Bellecour Lyon
reducing: external buildings and roads
increasing: sky view, tree border, size
Botanical Gardens Barcelona
reducing: fenced border
increasing: views, topography, tree borders, size
Emscher Park Duisburg
reducing: motorway, motorway noise
increasing: long views, topography, sheer size

Figure 14.5

PLANNING FOR CPULs: OPEN URBAN SPACE

Movement and housing . urban agriculture fields for Newark, *Bohn & Viljoen 1999*, see --> Practical Visioning

legend

◆ main entrance/exit

═══ major/minor feeding roads

─── main movement of which...

─── ...new routes/connections improving inner-city movement

scales □ 100m x 100m = 1ha
        □ 100m x 100m = 1ha
        □ 100m x 100m = 1ha

inner-city movement

The quality of Europe's open urban spaces is measured at the scale of pedestrians. Spaces that are easily accessible, easy to cross and/or inviting to sit, rest, wander are bursting with visitors, shoppers, traders and local residents (see variety of occupants/ occupation). Spaces not providing those characteristics don't burst.

Quality inner-city movement is strongly linked to pedestrians'/cyclists' access to individual pieces of land. As closer to the city centre an open urban space is, as more significant is its contribution to inner-city movement *(ie Place Bellecour, Old Plant Park)*.

A variety of access/exit points increases the potential for movement which still only happens if the open urban space allows safe, uninterrupted crossing *(ie Ronda del Mig, Mile End Park)*. Due to location or occupation constraints *(ie Thames Barrier Park, Old Plant Park)*, not all spaces have or require a large number of access points. Open urban spaces of particular function and therefore removed from urban routes *(ie Cemetry, allotment site)*, can even not at all contribute to inner-city movement.

Open urban space achieves a new unprecedented quality, when easily accessible pieces of land are inter-linked, thereby providing a continuous route through the city into the countryside *(ie CPUL)*.

economic return from ground-use

People's ability to generate income to live off is necessary and deeply satisfying. Using the ground, the earth itself, is one of the most archaic ways of generating that income, mainly in form of food production. It also allows the widely desired working in/with natural conditions.

Not using the ground itself can be related to the space's designation and its consequent layout and finish *(ie Place Bellecour, Ronda del Mig)*.

Drawing financial return from using the ground in a ritual way, happens on relatively large pieces of ground, but is an exception in Europe *(ie Cemetry)*.

Open urban space that has designed areas for agricultural production has gained extra occupational and spatial quality. Mostly, these open spaces contain allotment areas and generate that small income and huge satisfaction integral to allotment growing *(ie Old Plant Park, Mile End Park, allotment site)*.

Such partial usage of ground for actual production is imaginable in most garden- or parklike open spaces of some size without diminishing their spatial/visual character *(ie Thames Barrier Park)*.

A new, non-existing, landscape arises when the ground is worked to field-scale and commercialism. Then, significant economic return exists alongside the sustainable development and better protection of open urban space *(ie CPUL-urban agriculture)*.

Figure 14.6

FOOD IN SPACE

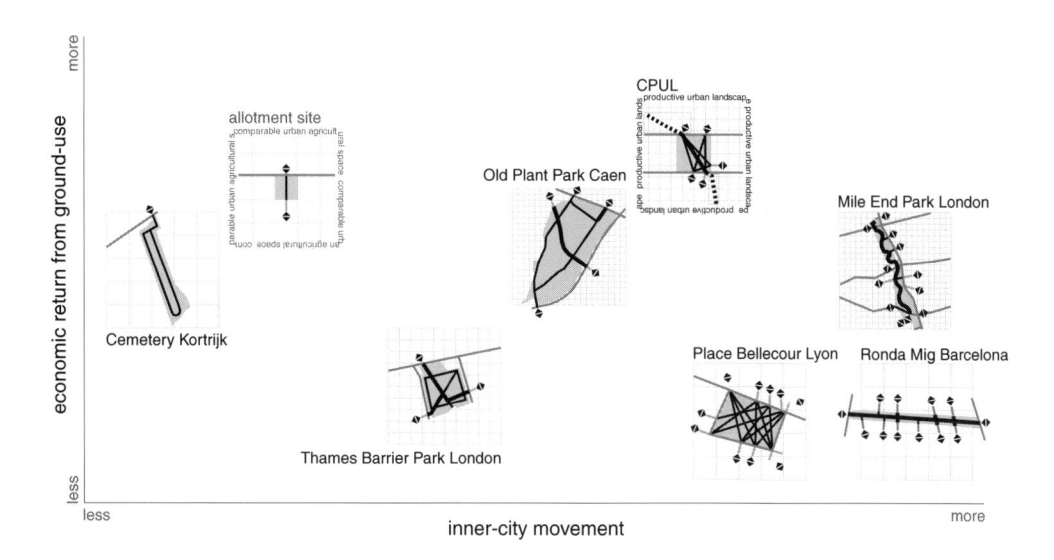

economic return from ground-use vs inner-city movement scatter plot showing: Cemetery Kortrijk, allotment site, Old Plant Park Caen, CPUL (productive urban landscape), Mile End Park London, Thames Barrier Park London, Place Bellecour Lyon, Ronda Mig Barcelona.

economic return from ground-use

Place Bellecour Lyon
indirect financial return from ground-use through events, visitors, traders, maintenance

Ronda del Mig Barcelona
indirect financial return from ground-use through visitors, traders, maintenance

Thames Barrier Park London
indirect financial return from ground-use through visitors, maintenance

Cemetry Kortrijk
fees from renting-out the ground/earth for ritual use employer

Mile End Park London
financial return from selling/trading grown produce (ecology park). employer

Old Plant Park Caen
financial return from selling/trading grown produce
financial benefit from composting, self-usage of grown produce

allotment site
financial return from selling/trading grown produce
financial benefit from composting, retaining seeds
self-usage of grown produce, lifestyle

CPUL
financial return from selling/trading grown produce
financial benefit from composting, retaining seeds
self-usage of grown produce, lifestyle and short-distance economy. local employer

inner-city movement

Cemetry Kortrijk
single entrance/exit, no connections

allotment site
few entrances/exits, non-permanent site access

Thames Barrier Park London
few entrances/exits, new connections between existing footpaths (S) and road (N), improved site access

Old Plant Park Caen
some entrances/exits, new connections between existing roads/paths, new residential routes, improved site access

CPUL
some entrances/exits, new connections between existing roads/paths and between different inner-city areas, improved site access

Place Bellecour Lyon
lots of entrances/exits, new connections between existing roads/routes and between different inner-city areas

Mile End Park London
lots of entrances/exits, new connections between existing and new roads/routes and between different inner-city areas, new pedestrian road bridge

Ronda del Mig Barcelona
lots of entrances/exits, new connections between existing routes and between different inner-city areas, new ways of moving for pedestrians and traffic

Figure 14.7

Horizontal and vertical fields . urban agriculture in Shoreditch, London, *Bohn & Viljoen 2000*, see --> Practical Visioning

legend

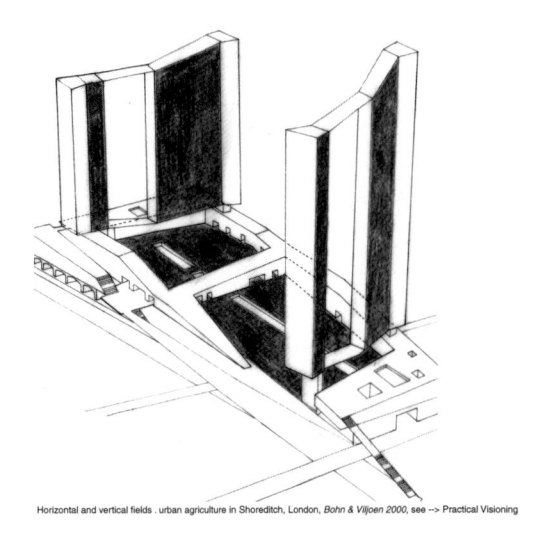

any planting of spatial importance

significant individual trees

water

scales 100m x 100m = 1ha
  100m x 100m = 1ha
  100m x 100m = 1ha

urban nature

To be in "nature" is desirable, and there is agreement that "nature" means living wilderness or countryside or forests etc., brief: non-urbanity. Urban nature is an attempt to measure the presence of such "natural" features in the urban environment. It describes an open space's potential for sensual experience involving, f.e., vision, touch, smell, taste, sound to create pleasure, friction or comfort.
Abundance of vegetation, both for visual and tactile experience, embodies most directly the presence of urban nature *(ie Botanical Gardens)*.
Water is also a strong visual and acoustic natural element of open urban space *(ie Parque de la Villette)* as are wide flat surfaces, both natural and sealed *(ie Place Bellecour, Emscher Park)*.
Agricultural vegetation will give open spaces an extra dimension by adding a greater variety of plants and forms of planting *(ie CPUL-urban agriculture)*.
Ecological planting exposes the occupants best to ecological cycles, the basis of all "nature" *(ie Emscher Park)*.
Open spaces that offer habitat to small wild animals are mostly either of larger size or committed to the protection of wildlife, which both encourages the flourish of nature *(ie community garden)*.
Lack of any of the above or dominance of built structures diminishes the feeling of exposure to nature *(ie Ronda del Mig, Five Squares)*.

persistent visual stimulation

Vision is the sense most used and therefore most addressed in contemporary European life.
Visual stimulation is crucial to maintain peoples interest, to challenge visitors. If a space does not satisfy people's vision, it will only be revisited for necessity rather than choice. To guarantee visual stimulation, movement, change are needed. Static is not stimulating, but even very little change is. As all spaces discussed here are *open* urban space, lighting and weather will continuously influence their optics.
The movement of people, cars, the flow of urban life is for many as stimulating as vegetation, esp. trees and flowers, changing over the seasons for others, though both are very different *(ie Ronda del Mig, Five Squares, community garden)*.
Staging various events in open urban space does not only persistently attracts different people, but reinvents the space's appearance again and again *(ie Place Bellecour, Parque de la Villette)*.
A combination of both, the inserted event and the every-day activity might be the most successful: seasons, traffic, leisure events etc. cater for a large variety of users and avoid mono-stimulation *(ie CPUL, Emscher Park)*.
Exhibitions held within open spaces add another layer of stimulation by making the space a second time visible *(ie Botanical Gardens, Parque de la Villette)*.

Figure 14.8

FOOD IN SPACE

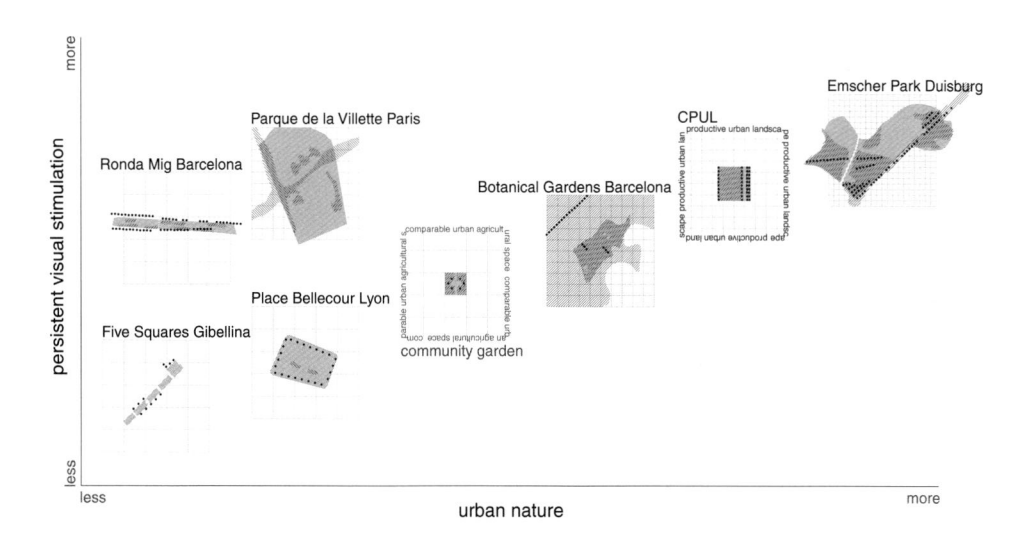

persistent visual stimulation (y-axis, less to more)
urban nature (x-axis, less to more)

Ronda Mig Barcelona
Parque de la Villette Paris
Botanical Gardens Barcelona
CPUL productive urban landscape
Emscher Park Duisburg
Five Squares Gibellina
Place Bellecour Lyon
comparable urban agricultural space configuration
community garden

persistent visual stimulation

Five Squares Gibellina
tree seasons. passing pedestrians and cyclists visitors. local events
Place Bellecour Lyon
tree seasons. passing pedestrians and cyclists visitors. local events
community garden
tree seasons. seasonal and changing vegetation passing pedestrians. visitors. workers. local events
Botanical Gardens Barcelona
tree seasons. seasonal and changing vegetation visitors. workers. local events. exhibitions
Ronda del Mig Barcelona
tree seasons. seasonal vegetation. passing pedestrians, cyclists and traffic. visitors. workers
Parque de la Villette Paris
tree seasons. seasonal vegetation. visitors. local and regional events. exhibitions
CPUL (urban agriculture)
tree seasons. seasonal and changing vegetation passing pedestrians and cyclists. visitors. workers
Emscher Park Duisburg
tree seasons. seasonal vegetation. passing pedestrians, cyclists and traffic. visitors. local and regional events

+ continuous change in lighting (day, night, sun...)
+ continuous change in weather (wind, rain, air...)

urban nature

Five Squares Gibellina
young individual trees
Ronda del Mig Barcelona
young individual trees. shrub and flower containers
Place Bellecour Lyon
mature individual trees. gravel. water
Parque de la Villette Paris
young individual trees. extensive lawn. flower containers. water
community garden
young (and/or mature) individual (fruit) trees. shrubs flower beds. lawn. (water). small wild animals
Botanical Gardens Barcelona
young individual trees. all varieties of shrubs, flowers, herbs, mosses etc. suitable to the existing climate. water
CPUL (urban agriculture)
young individual fruit trees and bushes. vegetable, herb, fruit, flower, grain beds or fields. lawn surfaces (water). gravel
Emscher Park Duisburg
young and mature individual trees. ecological planting. allotments. lawn surfaces. water. non-cultivated plants. gravel. small wild animals

+ sun, wind, rain, hot air, warm air, cold air...
+ insects
   for all open urban spaces

Figure 14.9

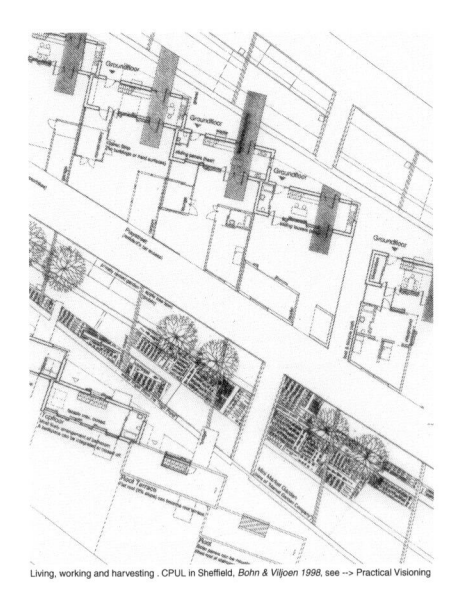

PLANNING FOR CPULs: OPEN URBAN SPACE

Living, working and harvesting . CPUL in Sheffield, *Bohn & Viljoen 1998*, see --> Practical Visioning

legend

low/medium/high biodiversity and soil quality

sources of major/ minor noise pollution

sources of calm

scales 100m x 100m = 1ha
        100m x 100m = 1ha
        100m x 100m = 1ha

environmental delight

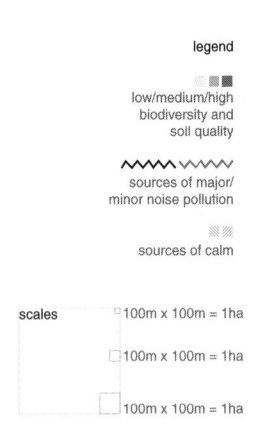

The desire for satisfying space, space to be content in, leads most people to open space of often very similar qualities. These qualities are grouped here as environmental delight. Apart from the existence of vegetation, water, weather and/or animals (see urban nature), people, in an environmental sense, seem to search calm, clean-aired, biodiverse and rich open spaces.

Open urban spaces provide both, acoustic and visual *calm* by distance from noise/traffic sources, mainly through the urban positioning of the space *(ie Five Squares, Mauerpark)* or through its sheer size *(ie Parque de la Villette)*. Vegetation, water surfaces and openness/horizon enhance visual calm (see sense of openness), both for people using a space temporarily *(ie Thames Barrier Park, Zoo)* and those living/working around it *(ie allotment site)*.

The co-relation between rich, healthy soil and biodiversity is expressed by combining them in one icon. *Biodiverse, rich-soil* open urban spaces are those holding abundant vegetation achieved either through spatially diverse planting *(ie allotment site)* or time-wise consecutive planting *(ie CPUL-urban agriculture)*. Provided there is distance from polluting sources, *clean air* is most likely found in open urban spaces encouraging wind (see size) and vegetation *(ie Cemetry, CPUL, Mauerpark)*.

variety of occupants

Inviting a variety of occupants is a quality of open urban space seeked by city councils as well as by users themselves for its potential to overcome societal boundaries, such as age, race, social status, gender, nationality, religion, culture etc.

Variety of occupants does not primarily register the numbers of users (though larger numbers often comprise more diverse users), but maps the differences between user groups. These are largely dependent on designation and, after that, design of the open urban space. Independent from the space's designation, members of any user group occupying a space with the same intention, will still belong to different backgrounds.

By offering single or multiple activities (see variety of occupation), the *designation* of a particular space encourages or restricts the diversity of its user groups for obvious reasons *(Parque de la Villette, Mauerpark, Cemetry)*.

Defining access to a particular designated space (see inner-city movement) by finance, therefore boundary, regulates the variety of occupants considerably *(ie Zoo, allotment site, Parque de la Villette)*.

The subtle, open, transgressing *design* of a space's non-financial boundary and its functional arrangement will determine its lasting attraction to a diverse usership *(ie CPUL, Five Squares, Thames Barrier Park)*.

Figure 14.10

FOOD IN SPACE

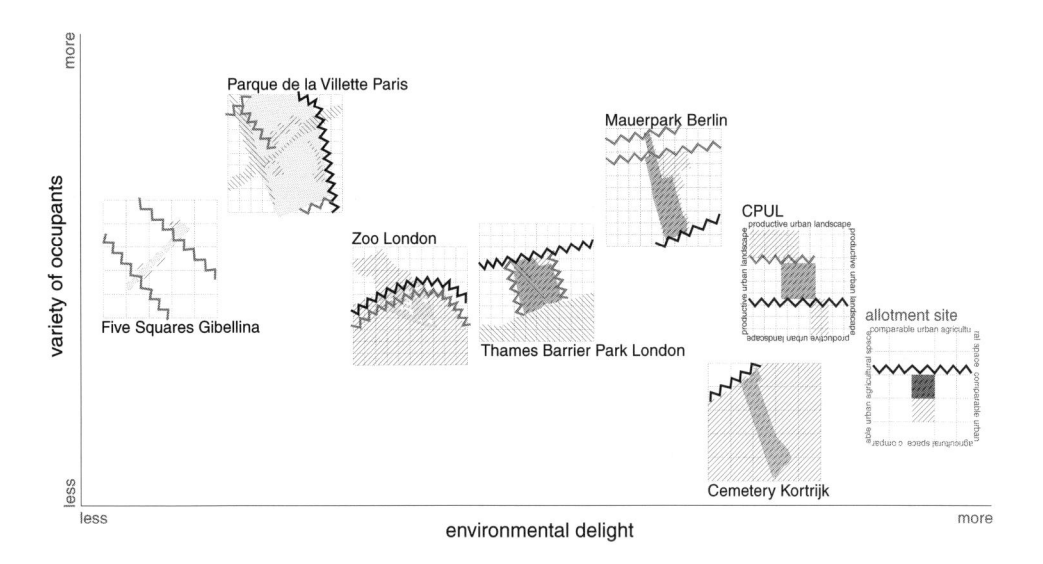

variety of occupants

Cemetery Kortrijk
mourners, workers. visitors
all ages, social & ethnic backgrounds. mainly locals

allotment site
allotment holders, passers-by. visitors/
all social and ethnic backgrounds. mainly locals
all ages, but mainly from 35 years on

Zoo London
private residents, workers. visitors
all social & ethnic backgrounds. local to international
all ages, but mainly children and parents/carers

Thames Barrier Park London
private & professional residents, passers-by. visitors
all ages, social & ethnic backgrounds. mainly locals

CPUL
private & professional residents, passers-by workers.
visitors. all social & ethnic backgrounds mainly locals.
all ages, but mainly adults

Five Squares Gibellina
private & professional residents, passers-by. visitors
all ages, social & ethnic backgrounds. mainly locals

Mauerpark Berlin
private & professional residents, passers-by. visitors
all ages, social & ethnic backgrounds. mainly locals

Parque de la Villette Paris
private & professional residents, passers-by
workers. visitors. all social & ethnic backgrounds
local to international. all ages, but mainly adults

environmental delight

Five Squares Gibellina
reducing the delight: lack of major vegetation
increasing the delight: quietness

Parque de la Villette Paris
reducing: noise/traffic connected to events,
amount of building on site
increasing: openness. water surface

Zoo London
reducing: major road traffic/noise, amount of
building on site. *increasing:* strong presence of
nature, both in vegetation and occupation (animals)

Thames Barrier Park London
reducing: road traffic/noise. *increasing:* quietness.
seasonally changing vegetation. water surface

Mauerpark Berlin
increasing: quietness. openness. abundance of
seasonally changing vegetation

Cemetery Kortrijk
increasing: quietness. openness. strong presence
of nature, both in vegetation and occupation (death)

CPUL
reducing: possible road traffic. *increasing:*
openness abundance of seasonally changing
vegetation. rich soil. visible ecological cycles

allotment site
increasing: quietness. abundance of seasonally
changing vegetation. rich soil. visible ecological
cycles

Figure 14.11

'nature' refers to the abundant presence of mostly uncultivated vegetation, of rain, sun, water, animal life, etc. perceived to happen mainly outside cities – in rural areas or the countryside. Seeming a contradiction, *urban* nature describes the presence of any of those characteristics in urban settings. Urban nature is used here to measure the potential of open urban space to satisfy, by design or designation, a human desire to be in 'nature'.

Whilst most contemporary spaces are concerned with similar urban issues, their main differences lie in the varied use of land, of the ground they sit on, of the local climate, vegetation or water, resulting in very different interfaces with the 'natural'. There is no qualitative judgement connected to those differences: open urban spaces withdraw for many reasons, mostly determined by their designation, from the 'natural'.

Nevertheless, the presence of urban nature as a quality is part of *environmental delight* which attempts to measure open spaces' success in environmental terms. It looks at qualities of vegetation, ground, air or sound and includes most people's desire for having sufficient access to the external, to being in the open (see Chapter 15). *Quietness*, both acoustic and visual, is an important measure of environmental delight, as the desire for quietness is one of the reasons for people to leave cities, whether on a holiday or weekend trip or to live outside them. *Noise*, its opposite, is becoming a more and more recognised environmental pollutant. Whilst there is different opinion about what noise is in detail, there is surprising consent about what it is in general. For open urban space, noise pollution is mostly traffic-, rather than industry-related, which might simply indicate strategies when positioning open space in the first place. *Air quality* (not mapped in the Open Urban Space Atlas) refers to fresh, clear, pollutant-free air registering little CO_2 and other gas emissions. Generally, good quality air is likely to be found away from polluting sources such as roads or industry. Open urban spaces have the potential to offer this condition, being often especially positioned or shaped so as to allow distance from such sources. As wind and the presence of vegetation help cleaning and/or dispersing polluted air, open urban spaces encouraging air movement and plant growth are also more likely to be clean. The *quality of vegetation and soil* is studied in the correlation of healthy soil (i.e. rich soil, free of pollutants and chemical fertiliser) and biodiversity. Assuming similar richness, soil free of chemical additives houses a greater diversity of plants and insects making it the basis for any long-lasting natural life in the first place.

About the open urban space atlas

The aim of the atlas diagrams is to explore the position of future CPULs amongst other contemporary open space designs and to extract their unique qualities. The comparison is based on our three criteria *spaciousness, occupation* and *ecology* with themes chosen to highlight the positive differences and qualities of CPULs compared to other open urban space.

The judgement of individual open spaces in such a format is rough, with the choice of criteria being 'sustainably subjective', but nevertheless the result indicates where the benefits of integrating CPULs into cities would lie.

Each diagram compares 6 (out of 14) contemporary European open urban spaces to a future CPUL and to either a community garden or allotment site as examples of open space types that hold urban agriculture already.

REFERENCES

Charbonneau, J. (1995). Three squares in Lyon. *Bauwelt*, **25**, 1428–1447. *(Place Bellecour, Lyon)*

Cheviakof, S. (2000). Barcelona Botanical Gardens. In *Ecological Architecture* pp. 56–59, Loft Publications. *(Botanical Garden, Barcelona)*

Engler, M. (2000). The garden in the machine: the Duisburg-Nord Landscape Park, Germany. *Land Forum*, **5**, 78–85. *(Landschaftspark, Duisburg)*

Ferrater, C. (2000). Jardin Botanico de Barcelona. In Carlos Ferrater pp. 250–263, *Editorial Munilla-Leria*. *(Botanical Garden, Barcelona)*

Gili, M. (ed.) (1997). Platform over the Ronda del Mig, Barcelona. *2G-Landscape Architecture*, **3**, 60–67. *(Ronda del Mig, Barcelona)*

Groupe Signes & Patel Taylor (2001). Thames Barrier, London, (United Kingdom). *A&V MONOGRAFIAS*, **91**, 90–93. *(Thames Barrier Park, London)*

Holden, R. (1996). Shaping open space. *Planning*, **147**, 8–11. *(Parque de la Villette, Paris)*

Holden, R. (2001). Greening the Docklands. *Landscape Architecture*, **10**, 82–87. *(Thames Barrier Park, London)*

London – The Photographic Atlas (2000). pp. 19–20, 128–9, HarperCollinsIllustrated. *(Mile End Park and Zoo, London)*

Lootsma, B. (1997). Redesign of the Schouwburgplein and Pathé multicinema, Rotterdam. *Domus*, **797**, 46–57. *(Schouwburgplein square, Rotterdam)*

Lubbers, P. and Wortmann, A. (2000). Hoog Kortrijk cemetery by Bernardo Secchi and Paola Vigano. *ARCHIS*, **10**, 30–33. *(Cemetery, Hoog-Kortrijk)*

Marquez, F. and Levene, R. (2002). Eduardo Arroyo: Plaza del Desierto, Baracaldo. In *In Progress 1999–2002* pp.154–169, El Croquis SL. *(Plaza del Desierto, Baracaldo)*

Perrault, D. (2001). Park in an Old Siderurgical Plant, Caen (France). *A&V MONOGRAFIAS*, **91**, 76–81. *(Old Plant Park, Caen)*

Pisani, M. (1994). Five Squares in Gibellina. *Industria Delle Construzione*, **273/274**, 4–15. *(Five Squares, Gibellina)*

Purini, F. and Thermes, L. (1992). New squares for Gibellina. *Deutsche Bauzeitschrift*, **7**, 1017–1022. *(Five Squares, Gibellina)*

Rossmann, A. (1996). Landschaftpark Duisburg-Nord – symbol of change in the Ruhrgebiet. *Bauwelt*, **37**, 2128–2135. *(Landschaftspark, Duisburg)*

Rousseau, S. (2000). Schouwburgplein: Mehr Bild als Platz? *Topos*, **33**, 18–24. *(Schouwburgplein square, Rotterdam)*

Uhrig, N. (1997). Mauerpark. In *Freiräume BERLIN* pp. 56–57, Callwey. *(Mauerpark, Berlin)*

Young, E. (2002). No walk in the park. *RIBA Journal*, **2**, 58–60. *(Mile End Park, London)*

15

DESIGNS ON THE PLOT: THE FUTURE FOR ALLOTMENTS IN URBAN LANDSCAPES

Professor David Crouch and Richard Wiltshire

The allotment garden is uniquely privileged as a form of urban agriculture and open space land use in Britain, in as much as local authorities have a statutory duty to provide land for present and future cultivators to grow their own fruit and vegetables. Moreover, the majority of allotment sites cannot be disposed of without explicit consent from central government, the granting of which depends on criteria relating to the supply and demand position for allotment land within the locality, not on the alternative designs authorities – or developers – might have on the land. This protection helps account for the continuing existence of these productive, peopled green spaces in urban landscapes throughout the country, despite the obvious commercial pressures to convert the land to other uses and realise its underlying monetary value. The original logic for providing and protecting this land was rooted in the poverty of manual workers over a century ago, however, and while allotments still provide for subsistence needs in many deprived communities, here as elsewhere the availability of cheap supermarket foods and the claims of work time and alternative leisure pursuits have undermined the revealed demand for allotments, which has declined markedly since the high point of the wartime 'Dig for Victory' campaigns. Not only is the land under pressure from commercial developers, but also from other green space users, jostling for room to accommodate those alternative recreations. The future for allotments depends on the creation of modern logics to justify their retention and on compromises between allotment cultures and alternative claimants on the land – as property but also as landscape and leisure resource. In this chapter, we explore the role that design might play in securing that new future – within rationales for protecting allotments and in mediating change.

Most of the key arguments in redefining allotments for the twenty-first century were aired in the 1998 Parliamentary Select Committee Inquiry into 'The Future for Allotments' (House of Commons, 1998), which discussed arguments about wholesome food in an era of deep concern over the quality and safety of diets, about healthy exercise and relaxation in the open air for the elderly, the discarded and the disturbed, about biodiversity and the support for urban nature which allotments provide, and about inclusive communities with a shared interest in a direct relationship with the land. It is notable, however, that design considerations received very little attention either in the Inquiry's report, or in the subsequent good practice guide *Growing in the Community*, which the present writers co-authored (Crouch, Sempik and Wiltshire, 2001), and which has a clear advocacy tone. The low profile afforded to design in discourses on allotments is a consequence in part of a chequered history of past encounters, as well as the difficulty of reconciling design concepts with the legitimate needs of allotment holders and the cultures they have spawned.

The design challenges that allotments present can best be understood by focusing on three distinctive contributions which allotments make to the urban scene: their role as part of green space (open or otherwise) in urban landscapes; their distinctive contribution to urban landscapes as sites of overt autonomous and creative activity; and their role as sites of negotiation – between land uses at the micro scale, but also between people from diverse backgrounds finding ways to reconnect as communities of propinquity and interest.

ALLOTMENTS AS (OPEN) GREEN SPACE

The few references that are made to design issues in *Growing in the Community* (Crouch et al., 2001) focus on the conflicting aesthetics of the public gaze and the spontaneous vernacular landscape of the allotment. Far from appealing to the onlooker as an open green space, allotments sometimes present a closed, ramshackle face to the world, an issue which underlies the citation as good practice in Kennet District Council's planning requirement for sites 'to avoid detrimental impact on landscape character and landscape features' and the need for self-built sheds to be subject to 'some degree of sympathetic regulation . . . to prevent the site from appearing too untidy . . .'.

The roots of the vernacular tradition in allotment structures, in making do with recycled and refashioned materials rather than splashing out on the ornate pinewood cabins so readily available in garden centres these days, lie in the origins of allotment gardening as a subsistence activity rather than a leisure pursuit, and in an inherent insecurity of tenure which is often compounded by planning blight, be it real or imagined. In rewriting the purpose of allotment gardens, *Growing in the Community* (Crouch et al., 2001) heaps praise on the use of recycled materials on allotments as a practical contribution to local sustainable development and the achievement of Agenda 21 objectives, a clear illustration of how aesthetic considerations can jar with other logics deployed to sustain the plot. At a deeper level however, the opportunity to self-build can be counted as part of the value of allotments to plotholders, a mode of self-expression beyond the reach of consumerist values and the instant TV designer garden, and a path to realising identity on an intimate, private scale, without reference to publicly authorised norms. For some observers this vernacular landscape has an enormous appeal of its own: 'It's the tumble-down bricolage that is utterly seductive' (Midgley, 2000). While this neatly captures the aesthetic sense required to appreciate the allotment vista, the fact that it is an aesthetic not always shared in the cultural mainstream helps explain the subsequent appearance of the quotation in *Private Eye* magazine's 'Pseud's Corner'.

The contested aesthetics of the allotment as viewed open space intersect with design considerations in the arena of planning policy guidance, and specifically in the current reformulation of PPG17 (Sport, Open Space and Recreation). The draft PPG17 calls for the creation of 'good quality' open space, and emphasises the need to 'apply design criteria in order to maintain or enhance the quality of the public realm' (Department of the Environment, Transport and the Regions, 2001, p. 22). There is scope for anxiety here, about the criteria to be employed to define 'well-designed' spaces (and their aesthetic assumptions), about who controls these criteria, and the sensitivity with which they may be applied.

A related and significant issue is the extent to which allotment sites, however visually seductive they might become, can actually be made into 'open' spaces, in as much as access conflicts with both the security of private property and the sense of ownership that are part of the value of the allotment to the plotholder. Although allotment land is often publicly owned, the crops and gardening paraphernalia required to cultivate them belong to the gardener, and are vulnerable to theft and vandalism. Complaints about inadequate fencing and unlocked gates are frequently heard, and throw the contradiction between openness and security into sharp relief.

Access isn't just a matter of interrupted views and freedom to gaze, however. Allotment tenancies establish rights of exclusive occupation, which can in practice last for decades, restricting the right of access to the gardening land itself to a fortunate band of early comers, particularly when individual plots are large. Through this process, allotments are transformed into semi-closed, semi-privatised public open spaces. Cultural expressions of the local 'ownership' that results can promote further exclusion, when young women, children, minorities, permaculturalists are made to feel unwelcome on the plots. Implicit in *Growing in the Community* (Crouch et al., 2001) is the need for good design at the site level to help overcome this unhelpful form of exclusivity, and realise latent demand for allotments, through simple measures like making plots smaller and sites safer for people of all ages and abilities. Fortunately, *Growing in the Community* (Crouch et al., 2001) does identify instances of good practice around the country of both local authorities and allotment associations working hard and in diverse ways to promote social inclusion on and around the allotment site – good practice which careful design could reinforce.

The role of design, in site layout and exposure to the public gaze, is to maximise the numbers who feel they have a positive stake in the survival of the allotments, as gardeners, observers or passers-by. Where exclusion persists, however, it is legitimate to inquire whether many of the benefits that are common to allotments and other forms of open space – such as healthy exercise in the open air, might not be more effectively delivered to larger numbers by assigning allotment land to other recreations: a threat to the plot from alternative green space uses. This is a strong rationale for the plotholders to take site design issues very seriously.

ALLOTMENTS AS URBAN LANDSCAPE

It can certainly be argued that allotments make a unique contribution to urban landscapes, a contribution which challenges conventional notions of both the urban and design. Urban allotments are a paradoxical echo of the countryside as it once was – a peopled landscape, yet accessible in the heart of the city, a space to construct the illusion of being in the 'countryside,' more friendly and accessible to people who wish to work with the land than the genuine contemporary countryside of agri-business, the social exclusivity of range-rover culture and the 'rural way of life' out in the green belt, the green lung that constricts the city's breathing. A peopled landscape, but one with a settled and quiet feel, a shared space with a hint of gathered silence. And, in common with other urban agricultures, allotments challenge conventional notions of town and country as productive *urban* growing spaces. This conflation of urban and rural has been a consistent source of difficulty for urban designers, because it contradicts imposed categories of what urban landscape is – or should be, expressed in particular notions of order and control, purity of form and clean edges. Yet it is just this contradiction of norms that people who are not professional designers appear to value (Crouch and Ward, 2002). Surrounded by urban Handsworth in the west of Birmingham, a plotholder named June captures a value in going to the allotment that feeds off its distinctive landscape: 'We come here for peace and quiet . . . It's like being in the country in the city. To come here on a nice summer's day you feel you're in the country' (Crouch, 2002). The regard felt by many people living, working, walking and cycling by allotments is testimony not only to their aesthetic value but also to the need to engage a wider population in discussion of the future of sites. Well-kept but chaotically landscaped sites provide a landscape otherwise unavailable, a diversity of human and biological culture that unsurprisingly means that they are valued amongst both ecological and cultural circles.

The allotment also provides an escape that the town park and other open spaces cannot deliver. However designed for informality, the park remains for recreation and is not a producing the landscape where people can grow, create and adapt their own ground, producing the landscape themselves. The organic growth and the feckless dynamism of the seasons and human attention provide relief and variety to the regularity even of row cultivation. And of course, the allotment space is both public and private (Crouch, 1998): individual plot holders invest in the ownership of their plot, but people who pass by or visit for a chat are part of its public display. Here is a landscape marrying regulation with disorder, an anarchic invention, a never-ending work in progress.

How at odds these values seem with Lord Rogers' brave new world of the (inedible) urban renaissance (Wiltshire, 1999). The report of the Urban Task Force (1999) reverts to the professionally self-serving model of an architecturally-led better city. Its diagnoses for obtaining good quality, high density urban living seek to revive a fifties style modernism that works from built form rather than human lives, communities and their actions and interactions. Such an approach would seem to fly in the face of the excellent work of community architects over three decades. High density living is assumed to be achievable through well-designed structures that respond to the built-form lessons of the middle of the last century. Where open space (including allotments) is acknowledged, it is afforded only a supporting role. Advocates for allotments must contest this reading of the city, and argue the evidence of value to human life and community identity that can emerge from the ways in which open, as well as interior space is used by individuals and groups. Exterior open spaces, like interior ones, offer the potential of shared and individual self-expression, involvement and commitment. Of course, the opposite can happen too, creating moments in which facilitators, including professional designers and community architects, can play an important mediating role. The lesson of the allotment landscape is that spaces can be enlivened through human activity; their use can engender feelings of ownership through investments of time and energy, as well as commitment. This approach to ownership both deviates from and complicates notions of ownership that refer only to legal or financial investment. Allotment landscapes offer one means of making high density urban living both humanely tolerable and compatible with the achievement of sustainable cities, environmentally, culturally and socially.

The contemporary friction between allotment landscapes and the new urban modernism echoes an earlier, failed attempt to give allotments a designer makeover, instigated by the Thorpe Report of the early 1970s and its advocacy of the 'leisure garden' concept (Thorpe, 1975), a designer-imposed solution to allotments in Britain. Borrowing ideas from some very successful models of allotment (leisure) garden design in other European countries, Thorpe's designer advisors were persuaded that these models would work in the UK. The report proposed spiral paths and 'fans' of plots and concrete blocks rather than the familiar individual sheds. In this respect these designs went further than those in other European countries, where the main format remains one of rectilinear plots in the familiar British style. The key characteristic of many sites across Europe is the presence of a large shed or chalet in which the law allows individual plotters to sleep. Wide trackways are provided on these sites in order to assist access and to become part of a more open design, although the landscape is softened by encroaching trees and plants that holders plant along the plot perimeter (Crouch and Ward, 2002).

Of course, the cost of redesigning sites on Thorpe's basis proved prohibitive, save for a few showcase projects. But there was another key reason why allotments à la Thorpe were never realised in Britain. Their culture was alien to the ways in which the culture of British allotments had developed. This is not to deny change, and the last three decades have shown that there is increasing diversity in the ways individuals want their allotments to look and the ways they want to use them. Moreover, there is little support anywhere for allotment spaces that are neglected. However, it is a mistake to confuse the loose-fit character of allotment landscapes with neglect. Redesign of allotments had been a persistent feature of manuals and government-inspired advice since the 1930s, and even before that garden city design included neatly arranged allotment plots, although on the traditional layout. But the gardeners persist in softening the edges, disrupting the lines, etching their culture into the landscape.

The lesson of these unhappy marriages between allotments and imposed design is that it would seem eminently sensible and practical to work from where people actually are in terms of using allotments. The application of design principles to allotments requires an acknowledgement of the way individuals work their plots and an understanding of the meaning allotment-holding bears for people within these gardening communities; in short, a respect for the landscape of allotments as an autonomous expression of culture alive in the city.

ALLOTMENTS AS NEGOTIATED COMMUNITIES

The landscape of allotments also plays a much valued role in the relationship between cultivation and identity building amongst plotholders. Allotment-holding is about much more than growing food:

> Working outdoors feels much better for your body somehow... more vigorous than day to day housework, more variety and stimulus.... Unexpected scents brought by breezes.... The air is always different and alerts the skin. Only when on your hands and knees do you notice insects and other small wonders. My allotment is of central importance in my life ... I feel strongly that everyone should have access to land, to establish a close relationship with the earth. Essential as our surroundings become more artificial....

> (Carol, Co. Durham, quoted in Crouch, 2002)

Working an allotment can engender community identity, through simple practices:

> If you give somebody anything, they say – where did you get it from – I say I grew it myself. You feel proud in yourself that you grows it.

> (Sylvester, Birmingham, ibid.)

The interchanges may be slight, but meaningful, and part of a process of confidence building too:

> We learn things from each other. You are very social and very kind. You make me feel good, you don't come and call at me ... things in your garden, little fruits ... which I have appreciated ever so much.

John, who is in his seventies, is talking with Allene, a woman in her fifties. She says:

> And I've learnt a lot from him. I've learnt ways of planting, I've learnt real skills about planting. I've learnt Jamaican ways of growing and cooking ... always said you could tell an Afro-Carribean allotment, an Asian allotment, an English allotment. And I've learnt about patience and goodness and religion too, it all links in.

They hardly ever speak together; they work quietly on adjacent plots, sometimes being there at the same time, but that does not matter. These kinds of expressions and exchanges describe the landscape of the allotment. It is a matter of understanding the landscape through human experience, relations and practices. Alone and together, plotters produce a landscape that is their expression of their culture.

However, this is not to say that allotment holders do exactly as they wish, although some do. Instead, working an allotment is composed of numerous negotiations. There are of course the rules that shape what can be done on a site; there are the negotiations between individuals who may not always be as agreeable as John and Allene, and which involve politics at a small scale. There is also a negotiation of the individual herself, working out, like Carol, her values and attitudes. In a very interesting expression of negotiated values and practices, the plotholders of Margery Lane site in the heart of the city of Durham found they had to negotiate very tactically between each other when their site was threatened (by its Diocesan owners) with closure and sale a decade ago. Some plotholders wanted to hold on to their dilapidated sheds. Others wanted to respond to the increasing public attention the site was given during the very public campaign for its retention (Crouch, 1994). Over a period of many months the plotholders held discussions and sometimes argued and eventually a few sheds were dismantled, but most were repaired and face-lifted. In consequence the site was saved, with the helpful support of the Durham Civic Society and the City Council.

THE PLOT FOR DESIGNERS

What then are the legitimate roles for designers from beyond the plotholding community? Our critique of the new modernism suggests a need for caution in exposing allotments to the ministrations of the architectural profession. There is a great deal of positive character inherited in the allotment landscape that deserves to be conserved, but this is a living heritage, best tended by its practitioners. Of course, it must be acknowledged that some sites have potential for recovery from neglect brought about through plotholder inaction, or landowner inattention, or even intended oversight (Crouch, 1998). There is a role here for the designer/architect, but it is a complicated one, requiring distinctive and discerning skills:

listening rather than telling; catalysing rather than directing. Crucially, this requires an empathy with the culture that produces allotment landscapes; the culture that the design expresses is largely given rather than imposed. Learning from the experience of good community architecture, designers can translate ideas using knowledge of, for example, efficiency of space use, tolerance, and the potential of particular materials. Plotholders may even have a grasp of the possibilities for articulating allotments with other open space and built uses, but where they do not, the designer's knowledge of how to realise the potential is of value. However, it is crucial to recognise that the knowledge that allotment holders use is lay knowledge, something that cannot readily be acquired by an unfamiliar designer, and which therefore provides a valuable resource with which to work, and which is spurned at the peril of repeating Thorpe's folly. The designer has to work with the landscape producers, the plotholders themselves.

We would argue that the proper role for the designer is in mediating change, in finding ways to accommodate the living landscape and culture of the allotment within the newer logics that *Growing in the Community* (Crouch et al., 2001) identifies as essential for their future; mediating access, for example, to realise the allotment as a more open and more widely valued green space, through designs which encourage and enhance the gaze, stimulating the viewer to ponder the merits of buying a fork and joining in, while protecting crops and property from misadventure with softened but appropriate security – the thorn behind the lowered wire. Within the site, design criteria might be integrated into new tenancy agreements, creating separate rules for gazed plots, or reduced rents linked to the retention of trees that add ornament at blossom time and opportunities in autumn for sharing the fruits. There are opportunities to exploit the subdivision of plots to accommodate busy people in innovative ways, mixing productive microspaces with shared facilities like benches and public art – things that reinforce the opportunities for social intimacy within the allotment and across its boundaries.

At a higher scale, the designer as urban planner can help integrate allotments in creative ways into the urban scene to achieve valuable synergistic effects. Crouch et al. (2001, p. 25) cite adjacent locations for plots and playgrounds as good practice – so that parents can cultivate in peace when the children get bored with worms and mud, while others without plots provide passive security when the gardeners are away. Co-location also creates opportunities for virement between the allotment and other green space uses, in accordance with changing demands, while other locational considerations can help bring allotments into the centre of a style of living, in harmony with the principles of sustainable development. Thus Hall and Ward (1998, pp. 206–7) call for the incorporation into new residential developments of: '. . . an allotment garden, which ideally would be provided in the communal open space in the middle of a superblock, entirely surrounded by houses and their own small private gardens. It would answer the insistent call for organic food from an increasingly sophisticated and worried public.'

What is essential, however, is that there be inclusion of the gardeners themselves in the design process, for both new and inherited allotment spaces, so that the outcome is compatible with the sense of ownership of the design, and of site and plots. There is a need, in other words, to work with the grain of allotment culture and the knowledge it embodies, to achieve better design through empowerment rather than imposition – design made sustainable through community pride, and where the details are left open, as opportunities for people to express their creativity informally at the microscale, where the arts of cultivation and the everyday accommodations of allotment life are played out.

REFERENCES

Crouch, D. (1994). Introduction to Crouch, D. and Ward, C., *The Allotment: Its Landscape and Culture*. 2nd Edn. Nottingham: Five Leaves.

Crouch, D. (1998). The street in popular geographical knowledge. In *Images of the Street* (N. Fyfe, ed.) London: Routledge.

Crouch, D. (2002). *The Art of Allotments*. Nottingham: Five Leaves.

Crouch, D., Sempik, J. and Wiltshire, R. (2001). *Growing in the Community: A Good Practice Guide for the Management of Allotments*. London: Local Government Association.

Crouch, D. and Ward, C. (2002). *The Allotment: Its Landscape and Culture*. Nottingham: Five Leaves.

Department of the Environment Transport and the Regions (2001). *Revision of Planning Policy Guidance Note (PPG) 17 Sport, Open Space and Recreation: Consultation Paper*. London: Department of the Environment Transport and the Regions.

Hall, P. and Ward, C. (1998). *Sociable Cities: The Legacy of Ebenezer Howard*. Chichester: John Wiley and Sons.

House of Commons, Environment Transport and Regional Affairs Committee (1998). *The Future for Allotments: Fifth Report of the House of Commons Environment, Transport and Regional Affairs Committee*. Volume I. HC560-I. London: The Stationery Office.

Midgley, S. (2000). Finding the plot. *The Guardian, Higher Education, 4 April*, p. 12.

Thorpe, H. (1975). The homely allotment: from rural dole to urban amenity: a neglected aspect of urban land use. *Geography*, **268**, 169–183.

Urban Task Force. (1999). *Towards an Urban Renaissance*. London: E&FN Spon.

Wiltshire, R. (1999). Towards an Inedible Urban Renaissance. *Allotment and Leisure Gardener*, **3**, 16.

PART FOUR

PLANNING FOR CPULs:
INTERNATIONAL EXPERIENCE

16

URBAN AGRICULTURE IN HAVANA: OPPORTUNITIES FOR THE FUTURE

Jorge Peña Díaz and Professor Phil Harris

This chapter focuses on the outstanding experience of urban and peri-urban agriculture (UPA) taking place in Cuba today. It has developed an innovative and comprehensive model supporting food production within the boundaries of its cities using a combination of top-down and bottom-up approaches.

Cuba, the largest island of the Antilles and the only remaining Western socialist society, has experienced two absolutely extreme situations in the last decades. On the one hand, the 1990s saw the most intense economical crisis of its history, produced after the collapse of the socialist block; on the other, it suffered the severe strengthening of the US blockade. The crisis generated a group of innovative responses; among them the first individual, spontaneous and later government-led support of UPA. The last few years have seen the successful, though partial, recovery of its main economic indicators and the maintenance of its social achievements. These improvements came accompanied by the introduction of market-oriented components into the socialist-planned Cuban economy. The overwhelming crisis that elicited the appearance of UPA has disappeared and new economic forces are now in operation. Meanwhile, the UPA implementation process itself has evolved towards more complex levels of organisation.

In this chapter the authors present the main characteristics of the urban agriculture currently present in Cuba, the causes of its origin, antecedents, people's and government's roles and other key factors emphasising the peculiarities of the Havana case study. They offer an overview of the main types of UPA production and producer partnerships to be found in Cuba, and of the integration of UPA into the planning policy of Havana, and comment on some of the conflicts that have appeared during its evolution. Finally, they discuss the most important lessons to be learnt from this experience and the factors affecting its replication.

THE CHALLENGE

Cuba, with a surface of 110 860 km^2, has a population of 11.22 million inhabitants and an increasing rate of urbanisation. More than 75 per cent of the population lives in cities. The collapse of the Eastern Europe block in the early 1990s, with which Cuba conducted over 80 per cent of its trade, resulted in massive economic dislocation and severe social impacts. In the same period, Cuba suffered the strengthening of the US 40-year-long economic, financial and political sanctions – the blockade. As a result of this combination, one of the most severely affected areas has been food supply: it is estimated that there was a 67 per cent reduction of food availability in 1994. A longstanding and systematic focus on social issues has resulted in particular achievements. For instance, the highest quality (and free) education and healthcare in Latin America are provided to all Cuban citizens. These actions have been reflected in other areas such as science, sports and culture. For example, despite having only 2 per cent of Latin America's population, Cuba has 11 per cent of its scientists and has developed an advanced network of world class research institutes. Female participation in society is outstanding and Cuba is ranked 58 in the UNDP Human Development Index 2000.

Since the 1820s Cuban agriculture has been dominated by sugar production and in the 1860s Cuba became the world's largest sugar exporter. After the revolution in 1959, and its agrarian reforms, sugar continued to play a dominant role and most agriculture involved extensive mechanised cultivation of 'exotic' species and other export crops. Much work was done to develop the National Agricultural sector into a highly mechanised one, with intensive use of agro-chemical products. Non-sugar agriculture had also largely been based upon large state farms and co-operatives. Cuba's agriculture and food industries were heavily dependent on imports, including 100 per cent of wheat, 90 per cent of beans and 57 per cent of all calories consumed. The agricultural sector imported 48 per cent of fertilisers and 82 per cent of pesticides (Rosset and Benjamin, 1994). Balanced animal feeds were largely based on maize and other cereals, of which 97 per cent were imported (Deere, 1992, Rosset and Benjamin, 1994). From 1989 these imports were drastically reduced and agriculture faced an immediate crisis: imports of wheat and other grains for human consumption dropped by more than 50 per cent (Deere, 1992) while other foodstuffs, with the exception of powdered milk, declined even more, with a drop of more than 80 per cent in the availability of fertilisers and pesticides (Rosset and Benjamin, 1994). Electricity generation was drastically reduced, affecting food storage capacities; fuel and parts for maintaining transport vehicles, oil and other inputs became scarce commodities, driving the traditional agricultural system into critical disruption, with a subsequent impact on the availability of nationally produced food, additionally affected by the reduction of food imports.

Together with achieving general economic recovery, the challenge was then established to generate effective mechanisms to meet people's food needs and to fulfil the government's commitment on this social issue. Taking into consideration the scale and coverage of the crisis what strategy would serve to meet the food demands of a predominantly urban population heavily dependent on the produce of a depressed countryside?

PEOPLE'S AND GOVERNMENT'S RESPONSE

After the crisis, Cuba lost more than 75 per cent of its import and export capacity. In response, a nationwide package of emergency measures was established within a process known as Período Especial en Tiempo de Paz (Special Period). To confront the problems, which included reduced food supplies and agricultural inputs, the government instigated a programme of reforms which included the redirection of its trade towards the world market and the introduction of some market style reforms in the domestic economy. Additionally, the process included the adoption of innovative approaches directed towards the downsizing of the central government and many industries; decentralisation and reorganisation strategies; import substitution and alternative approaches in many areas among other measures.

As part of these reforms a national alternative agricultural model (NAAM) has been developed since 1990. One important aspect of this model is the replacement of high levels of imported agricultural inputs with alternative, indigenously developed methods for pest and disease controls, soil fertility and other innovative issues. Other aspects have included the restructuring of the land property pattern of the large state-owned farms into smaller units with co-operative property, and the permitting of a free market[1] for foodstuffs. The crisis had generated the immediate individual response of many citizens and groups and in parallel encouragement was given to individuals and/or co-operatives in urban and peri-urban areas to become vegetable producers. Several forms of exploitation of the open spaces within the urban frame but also of available areas within productive and services, educational, recreational, and healthcare facilities were established. In the period 1990–1994 it was estimated that more than 25 000 people were linked to about 1800 hectares of so-called popular orchards. These were set up spontaneously by individuals on open land in response to the food crisis. The present experience of urban food production has evolved

---

[1] Similar experiences were carried out during the 1980s allowing the selling of agricultural produces. Several factors contributed to the termination of these experiments. In this sense the free market is rather a saga of previous attempts.

towards an organised and co-ordinated programme within this process. UPA should be seen as a component of a wider strategic process rather than as an isolated one.

RURAL AGRICULTURE AND URBAN HORTICULTURE IN CUBA

The ultimate goal of NAAM is to develop less mechanised, more labour intensive operations which involve local populations in low external input sustainable food production enterprises. An important feature of Cuba's ability to respond rapidly with an alternative model has been its success in mobilising science and technology and the social investment in education. Cuba had much earlier made a decision to promote scientific knowledge-based development, for example, in agricultural biotechnology and biological control. Many of the 'new' technologies had been actively researched, specifically for the Cuban environment, for a decade or so before circumstances made their implementation imperative. This made it much easier for the implementation of the alternative model, avoiding problems of technology transfer or reception that might be experienced elsewhere. At the research level there is support for the principles of organic production. There are very good links between the research and extension organisations and consequently a very short time lag between research innovation and field application of research results.

The use of bio fertilisers, for instance, is advanced by any standards. Sufficient *Rhizobium* inoculants for leguminous crops for the whole country are produced, and free living nitrogen fixing organisms such as *Azotobacter*, and phosphorus solubilising microorganisms such as *Bacillus*, are being mass produced in government-run, low technology production units. Vesicular arbuscular mycorrhizal inoculants, fungi which colonise the roots of plants and help them obtain nutrients from the soil, are also being mass produced and distributed. Maintenance of soil fertility on even large farms now depends significantly on a range of management practices, including reduced tillage, the adoption of crop rotations with the incorporation of green manure such as *Sesbania*, intercropping, and low input rotational grazing systems to build soil fertility on dairy farms. Organic inputs include livestock manure, other agricultural wastes and compost produced in industrial-scale worm composting operations. For instance, Cuba in 1992 produced 93 000 tonnes of worm compost.

[2] Until 1998 Mario Gouzalez Novo was head of the Havana Ministry of Agriculture, Urban Agriculture Department.

Within this alternative model urban and peri-urban agriculture was quickly identified as a viable large-scale response aimed first and foremostly at the partial recovery of the threatened food security. Since 1989 hundreds of state-owned hectares of land were authorised to be used by citizens willing to cultivate them. The Ministry of Agriculture structured itself to provide citizens with advice, material support and encouragement in the management of agricultural activities. The movement was of such a magnitude that in 1993 the Ministry of Agriculture together with city's government, in response, organised an official structure to pay attention to the phenomenon covering the provincial, the municipal and the local levels (Cruz, 2001). Several national NGO's with varied platforms were also integrated and some International ones and other organisations supported the movements as well.

As a recent phenomena in the Cuban context (at least in its newer experience), several definitions for UPA have been offered by researchers and different stake-holders. González Novo[2] and Murphy (1999) define urban agriculture in Cuba as 'an intensive, high input (organic pesticides and organic manure) and high output system favouring the production of a

DRUMFLOWERS study by Tom Phillips
CIENFUEGOS

Plate 1

Plate 2

CUATRO CAMINOS
CIENFUEGOS

Plate 3

PASTORITA
CIENFUEGOS

Plate 4

PUEBLO GRIFO NUEVO
CIENFUEGOS

PUEBLO GRIFO NUEVO
CIENFUEGOS

Plate 5

[3] Former Vice President of the National Group for Urban Agriculture.

diversity of crops and animals during the year.' He describes the main idea of urban agriculture in Havana as 'Production in the community, by the community, for the community,' adding that 'urban agriculture is very much seen as a way to bring producers and consumers closer together in order to achieve a steady supply of fresh, healthy, and varied products directly from the production site to the consumers.' The latter catches the principle of connecting most of the stakeholders at the community level as a key component to ensuring the success of agricultural production. Meanwhile Peña Díaz[3] (2001) asserts 'urban agriculture is the intensive production of food by using organic means within the perimeter of the human settlements and its periphery, based on the maximum use of the productive potential of each territory. It includes the use of the available land, the high cultural level working force, the interrelation of crop–animals and the advantages of the urban infrastructure. It encourages the diversification of crops and animals guaranteeing its stepped permanence all year round.' Other definitions provide more elements, however a pragmatic concept generally in use asserts that 'in Cuba urban agriculture is that occurring 10 km around the provincial capitals, 5 km around the municipal main towns, and 2 km around the smaller towns' (Guevara Núñez, 2001). Average people, on the other hand, just identify urban agriculture with the best known form of production, the organopónicos (popular organic orchards). Most of these definitions fail to address some of the key issues relevant for an adequate understanding of the activity, such as location, stage, activities, destiny, etc. Most of them seem in a way normative since they provide a vision of how UPA should be, i.e. an aspiration, or they are too pragmatic and too vague as to give a clear understanding of the urban character of the processes and its implications. Issues such as differentiation of intra-urban agriculture from peri-urban agriculture, or the identification of rural agriculture within the urban or peri-urban sectors of the city have been scarcely, if not at all, discussed in the Cuban context.

Beyond the different definitions the Cuban model for urban food production has been implemented nationwide. This is a key difference to other countries since the same techniques, organisation pattern, pest and diseases control products can be found all over the country. The elements characterising the model are then the adoption of a set of production forms and the assimilation of current association forms belonging to the rural agriculture, the support of the scientific networks linked to the Ministry of Agriculture, the commitment of the Government structures at all levels to support it and the adoption of a structure with strong influence at the grass-root level. This National programme is presently co-ordinated by the National Group for Urban Agriculture (GNAU) which is made up of representatives of scientific institutions of the Agriculture Ministry and other stakeholders. It conducts the promotion, implementation and evaluation work of 26 so-called subprogrammes. Each subprogramme is dedicated to one specific type of production or aspect of the industry. For example, there is a flowers subprogramme, a vegetable production sub-programme, a herbs and spices subprogramme, an environmental issues subprogramme amongst others. The national group evaluates the performance of each municipality according to the implementation of the different subprogrammes and their achievements. The result of this evaluation is the selection of the most outstanding municipalities, and the identification of best practice cases. All of the country's 169 municipalities have developed urban agriculture within this programme.

There are several forms of production corresponding to the types of agricultural techniques used. Each of them has varied efficiency indicators (yields) ranging from 8 $kg/m^2.yr$ to 25 $kg/m^2.yr$.

[4] UBPC: Unidades Básicas de Producción Cooperativa – Basic Units for Co-Operative Production; CCS: Cooperativas de Créditos y Servicios – Services and Loans Co-operatives; CPA: Cooperativas de Producción Agropecuaria – Agricultural Production Co-operatives.

Other important components are the production supporting entities, which also belong to the sub-programmes though they do not directly produce food. They provide consultancy and advice services. Producers are grouped in different kinds of partnerships, which have been established with specific payment, commercialisation and taxation regimes, ranging from self payment to salaries; in-situ commercialisation to tourist facilities supply and usufruct to tax on land use respectively. These forms of association largely reproduce the structures present in rural agriculture.

Main forms of production are the popular organic orchards (so-called organopónicos) and the high efficiency organic orchard (organopónico de alto rendimiento, OAR), the popular and intensive orchards, state owned self-production areas (auto-consumos estatales), popular orchards divided into parcels and intensive orchards (huertos populares parcelas y huertos intensivos), and covered houses (casas de cultivo). The production supporter entities are organic material production centres (centros de producción de materia orgánica), agricultural clinics and shops (consultorios agrícolas, afterwards tiendas consultorio agropecuario), and nursery houses (casas de posturas). Farmers are associated in structures similar to those of the rural agriculture, in different types of co-operatives such as UBPC, CCS, and CPA[4]. The unplanned creation of Farmers Clubs by urban dwellers was notable for the way it supported the wider idea of UA policy.

URBAN AND PERI-URBAN AGRICULTURE IN HAVANA

The capital city Havana covers an area of 727 km², 0.67 per cent of the total area of Cuba. Its population growth is 1.8 per cent per year and it has about 2.2 million inhabitants, which make up 20 per cent of the Cuban population and 27 per cent of its urban population (DPPFA, 2000, González Novo and Murphy, 1999). The average population density is 3014 persons/km². In spite of the decentralisation policy that took place during the previous 40 years in favour of the inner provinces, Havana's physical extension, its services and industrial infrastructure increased by threefold (Palet, 1995, DPPFA, 2000). In the year 2000 Havana still held 34 per cent of National industrial production, the main administrative and political departments as well as most of the specialised healthcare, educational, and cultural services. Additionally, Havana is the centre of scientific activity and one of the most important foreign tourist attractions (DPPFA, 2000). The crises of the 1990s affected the employment indicators, though the recovery has brought new labour demands. There are also difficulties in Havana with the water supply infrastructure.

Agriculture in Havana, for example, is backed by an extension service that would be the envy of most of the world. The Ministry of Agriculture's Urban Agriculture Department has a large number of extension officers to provide free extension and training services to the organic farming sector, visiting regularly, and encouraging and advising on the use of organic techniques.

HAVANA: ANTECEDENTS AND CURRENT DEVELOPMENT

The presence of Chinese farmers in the outskirts of Havana during the first decades of the twentieth century is still part of Cubans' collective memory. This practice declined probably as a consequence of the reduction of the Chinese community. It shows however that any type of UPA in Havana has had an alternate presence, especially during the last 40 years. In this period it has always appeared as a response to extraordinary situations: in the 1960s in the inspiring atmosphere that the radical transformations of Cuban society brought to the capital. Havana's green belt was proposed in the 1964 Master Plan. It was envisaged as a way of overcoming the capital's dependence on food provisions from the inner territories. After an effervescent period of activity it decayed. Later, in the nineties, in the face of major economic and financial shortcomings, UPA appeared again amid a group of innovative measures within the 'Período Especial.' In the capital many had started to grow vegetables in the backyards, a practice with a long tradition, and even in the parterres in front of their houses. When the national and provincial governments encouraged people to occupy every single free space with crops, vegetables were planted even in front of the Ministry of Agriculture and in the backyard of the state council building.

In the early 1960s the proposal of creating El Cordón de La Habana (Havana's green belt) represented a noticeable change in the infrastructure of agriculture, in the improvement of the farmers' lifestyle, as well as an increase in the supply of food to the population of the capital. An ideological conception envisaged the city as a parasite of the countryside (Segre et al., 1997). The objective was to provide the city with a productive surface that would empower the self-feeding capacity, but also provide it with a number of recreational areas for the urban population of the largest city on the island. The plan conceived of a Fruit trees belt on the land closest to Havana, followed by a milk producer belt, providing it with additional environmental commodities, and health assurances, achieving a more efficient use of lands (most of them were abandoned plots or were previously used with speculation purposes) through the integration of the public and the private sectors (Ponce de Leon, 1986). Around 21 566 hectares of land were put on an exploitation regime, 5032 hectares of them for pasture and 16 533 hectares for fruit trees, citric trees, coffee, gandul, and forestry. Forestry areas were conceived as micro-forests and parks, such as the (National) Botanical Garden, (National) Zoo, Metropolitan Park, (the so-called) Solidarity Park, and the Lenin Park. Several tree-nursery areas were supposed to produce 100 million units necessary to plant the belt. It also included: 'the construction of 80 small-size dams and micro-dams for irrigation, more than 180 km of agricultural roads and paths, around 1000 houses and 8 small villages for the farmers and workers of the territory, the terracing of the hills for coffee plantations and the planting of tree-curtains against the wind.' (Ponce de León 2000).

The plan was carried out, and between 1967 and 1969 millions of nursery plants of different species were planted. 41 per cent of these plantations remained 20 years later. In 1986 it represented around 2416 hectares of fruit tree plantations located in the eastern part of the province, composed mainly of mango trees, 940 hectares of coffee to the south and southwest zones, 268 hecaters of citric trees, 3019 hectares of pasture and more than 1721 hectares of forestry. 52 micro-dams and the Ejército Rebelde dam with a capacity of 52 000 000 m^3 to be used for irrigation and leisure purposes, were also built. 2850 houses were built between 1967 and 1973, both individual ones and ones in the 12 residential communities. Also, around 200 km of roads were built and there are still remainders of the Tree-Curtains against the wind (Ponce de Leon, 1986).

THE EXPERIENCE OF THE 1990s

In 1991, through a strong information campaign, Havana's city government encouraged the city's population to use every single piece of open space in order to produce food for direct consumption. This generated an immediate response and several forms of exploitation of the open spaces within the urban frame but also of available areas within productive, service, educational, recreational, and healthcare facilities were established. Most of them were areas smaller than 1500 m² which were given for a temporary but undefined period of time, to be exploited by one or various families, neighbours, students and docent personnel, workers of factories, etc. Initially these areas were used to grow vegetables and roots, and to keep poultry and smaller cattle. A massive breeding of pigs in these areas was illegally extended to central urban areas including houses and flats. This process has evolved towards better organised forms of productions.

The entire territory of the capital is considered urban by the Physical Planning Office (PhP) and by extension all agricultural production within Havana is considered Urban by the GNAU. The most recently approved master plan for the city (Territorial Ordering Scheme) considers, for the first time ever, UPA as a permanent urban function. It is a tacit recognition of the importance given to UPA, by the institution. However it contains some reactive position towards UPA since, for example, it prohibits its occurrence in the most central areas.

The Metropolitan Horticulture Enterprise Havana, established by the Ministry of Agriculture, runs most of the production of vegetables and herbs in Havana. It controls and supplies most of the production and management of UPA in the province. It additionally facilitates tax payments for individual and co-operative producers.

[5] The two most compact municipalities, Old Havana and Centro Habana do not have urban agriculture production.

There is one provincial delegation of the Ministry of Agriculture in Havana. It has a three-level structure and is presided over by one delegate. 15 municipal delegates preside on the equivalent structure in the municipalities and there is one grass-root level delegate known as 'extensionista' in the districts. The delegates at the provincial and the municipal levels belong to the respective government assemblies in such a way that they can take part in the decision making process. There are 13 urban farms in 13 municipalities[5] which belong to the entrepreneurial branch but establish close co-operation with the lowest level of the delegation. Additionally, they closely co-operate with the government institution at that level, the so-called Consejo Popular. This is a major opportunity to enhance the participatory approach they have selected for their work.

Main conflicts with UPA in Havana are related to the usage of water, the maintenance of soil fertility, and its insertion into closed cycles. Water scarcity is not a major problem but problems with the ageing infrastructure allow more than 50 per cent of water pumped for human consumption to leak. Several districts are permanently condemned to water shortages. In contrast, the main source of water for irrigation used by UPA is tap water, highlighting a conflict that is far from being solved. Soil fertility due to the organic character of the activity is based on the importation of organic matter from remote areas and the production of compost. It is maintained with inputs of animal manure, mulching, and composting. Composting is either by static heaps or by worm composting. Soil fertility can be maintained or enhanced by increasing inputs and continuing with intensive production or by adopting less intensive production. The prospect of reducing the intensity of production in a system designed to provide a valuable addition to food supply at a time of economic crisis, is unattractive; increasing the organic input is the favoured option. This results in difficulties since access to and availability of wastes is limited by scarcity of resources. Furthermore, UPA has developed without the restrictions of an established market in land speculation. Many popular organic orchards (organopónicos) for instance occupy plots estimated to have a very high value if sold on the open market. The development of such a market could affect UPA development if additional measures, e.g. integration into planning policy, are not further developed.

The capital has developed all of the so-called UPA subprogrammes and Havana's organic urban agriculture is extraordinarily successful, such that fresh produce offered at the farm gate is cheaper than that brought from the countryside into the city's free markets. Actually, in contrast to other territories where the focus of the development of UPA has been placed on the provision of jobs, in Havana it looks to counter the range of prices for food related products. Havana is an extraordinary case where organic vegetables are actually the cheapest option. The production of the UPA has been linked to social institutions such as kindergartens, hospitals and schools whose restaurants receive fresh organic produce on a daily basis. It is part of a strong information campaign promoting the intake of vegetables as a positive nutrition practice and attempts to counter traditional customs which do not favour vegetables intake. The campaign includes TV-broadcast lessons on how to prepare vegetable-based meals and on their nutritional value, and programmes to advise farmers on techniques and methods for cultivation. Additionally, the Junior High Education system has implemented a new subject called urban agriculture. It shows young students what UPA is in theory and practice. Often students even cultivate some crops in the school's backyard and gardens.

CONCLUSIONS

In Cuba the urban agriculture model implemented matches the crisis model described by Deelstra and other authors. The driving forces were the economical difficulties derived from the drastic modification of economic relationships. After the crisis an alternative model was pursued and within it urban agriculture was identified early on as a means to ensure food availability for a large number of urban households and has recently been identified as an important provider of jobs. Support for food related issues had prevailed as a prioritised issue in the Cuban political agenda since the early 1960s and this experience seems to be inscribed also in UPA.

Presently, more than ten years since the initial implementation of urban agriculture, the process has evolved into one of highly complex structures and relationships. A strong emphasis has been placed on the socioeconomic dimension, for instance, food production and its nutritional values; in second place, the potential of UPA as a job provider and its influence in the lowering of food prices have been recognised. Environmentally, its organic character is its main asset but other aspects have been overlooked or neglected. Largely (at least initially) due to the lack of hard currency to purchase fertilisers and pesticides, many producers individually started to use derelict urban spaces and are now using organic production methods, which has become a requirement for most urban producers. There is also a growing awareness of environmental damage caused by intensive conventional techniques. These statements show that the government is eager to promote urban and organic food production firstly for both economic and social reason, but also for environmental reasons.

In terms of its socio-economic situation and political system, Cuba is unique. Few developing countries have invested so much in their human capital, nor have they achieved similar social indicators. Probably none has suffered such stress on its economy as that generated in the early 1990s, nor the effects of current economic and financial sanctions. In part, the problems Cuba now experiences are the result of an economy pushed to be dependent until 1990 on few trading partners, and of an agricultural strategy in which export cash crops dominated, food security was not a priority, and food self-sufficiency was poorly covered. Undoubtedly, part of the success in adopting an alternative agricultural model can be ascribed to the central influence on markets, and indeed on all aspects of the economy and society in general, of central government. In spite of its uniqueness there are important lessons to be learnt from the Cuban experience.

What makes it outstanding is firstly its comprehensiveness. Secondly it is a matter of scale. In a very short period Cuba has observed a tremendous increase in the volume of production, in the areas dedicated to urban production and likely a variation in the nutrition habits of some sectors of the population favouring an increase in their daily consumption of fresh vegetables. González Novo and Murphy (2001) have described the Cuban experience as the 'world's first nationwide co-ordinated urban agriculture programme, integrating access to land, extension services, research and technology development, new supply stores for small farmers and new marketing schemes and organisation of selling points for urban producers.'

There has been a close, effective association between the development of agriculture, local government and local democracy. The participation developed at the community level makes it a successful combination of top-down and bottom-up approaches. The short period, and the scale, of its implementation are surprising. Earlier investment in human capital and specifically in locally-oriented agricultural research and development skills and infrastructure have paid dividends and the close links between research organisations and the excellent extension services have delivered appropriate and timely research outputs. A mixture of innovation and determination are today making the Cuban experience a success story. Necessity has persuaded Cuba to adopt a largely organic approach to agriculture in which urban and peri-urban agriculture occupy a relevant role. Together with this organic approach, the presence of a steady political will to ensure food availability for all seems to have had a strong influence. The existence of political willingness seems to be a key factor in ensuring the development of this kind of experience even though the realisation of all advantages is limited. For instance, a strong emphasis initially on food production and later on jobs generation, nutrition quality and education issues to the detriment of environmental considerations beyond UPA's essentially organic character.

Havana represents an interesting case study since it has experienced most of the contradictions associated with this kind of urban industry. The level of integration into Urban Planning Policy is distinctive and has been achieved though a more flexible and proactive approach resulting from a better understanding of the potential positive role and risks associated with UPA and its potential contribution to comprehensive policies.

It remains to be seen to what extent Cuba's organic experiment will survive when the national economy recovers, when the prospect of agricultural input returns, and when the full insertion of Cuba's economy into the world market becomes a reality, including the influence of the American agri-business. However, urban agriculture occupies a distinctive position within this process and it seems to be more likely to survive these influences even if the extension of the alternative model is reduced to small spots or even disappears. In the meantime lessons are being learnt which could have major implications worldwide.

REFERENCES

Cruz, M. C. (2001). Agricultura y Ciudad una clave para la sustentabilidad, Fundación de la Naturaleza y el Hombre, Havana.

Deere, C. D. (1992). *Socialism on one island?: Cuba's National Food Program and its prospects for food security*. Institute of Social Studies. The Hague.

DPPFA (2000). Esquema de Ordenamiento Territorial 2001 (Master Plan). Dirección Provincial de Planificación Física. La Habana.

Dresher, A., Jacobi, P. W. and Amend, J. (May 2000). *Urban Agriculture: Justification and Planning guidelines*. http://www.city farmer.org/uajustification.html

González Novo, M. and Murphy, C. (1999). Urban Agriculture in the city of Havana a popular response to a crisis. In *Growing Cities, Growing Food Urban Agriculture on the Policy Agenda*. A Reader on Urban Agriculture. Deutsche Stiftung fuer internationale Entwicklung (DSE), 329–348.

Guevara Núñez, O. (2001). Demostración de que si se puede. In *Granma*, (official newspaper of the Cuban Communist Party) 1 February, 2001, p. 8.

Palet, M. (1995). Estructura de los asentamientos humanos en Cuba. Doctorate Thesis. Institute of Tropical Geography, Havana. (Not published)

Peña Díaz, J. (2001). *The integration of urban and periurban agriculture into the planning policy of Havana*. Master of Science Thesis, Department of Infrastructure and Planning NR 01–169, Royal Institute of Technology, Stockholm.

Ponce de Leon, E. (1986). El sistema de áreas verdes de la Habana. Primera Jornada Científica del Instituto de Planificación Física. IPF. La Habana.

Ponce de Leon, E. (2000). Personal interview at the Grupo para el desarrollo Integral de la Capital. September 2000. Havana.

Rosset, P. and Benjamin, M. (1994). Two steps back, one step forward: Cuba's National Policy for Alternative Agriculture, Gatekeeper series No. 46.

Segre R., Coyula, M. and Scarpaci, J. (1997). *Havana: two faces of the Antillean metropolis*. World cities series. Chichester.

17

CUBA:
LABORATORY FOR URBAN
AGRICULTURE

André Viljoen and Joe Howe

The extent, infrastructural support for, and reliance placed upon urban agriculture in Cuba, means that it provides a rich source of information for the study of urban agriculture (Bourque and Canizares, 2000), (Caridad Cruz and Sanchez Medina, 2003). As such it may be considered a laboratory in which to observe and speculate upon the future shape of productive urban landscapes.

As Harris and Penna's chapter makes clear, urban agriculture was introduced into Cuba as a matter of necessity (see Chapter 16). What one sees in Cuba is a system pragmatically placed with little time to develop design strategies that can respond to urban agriculture. Detailed observations, which can be made in Cuba, help us to understand what a Continuous Productive Urban Landscape could be like. The following section draws on observations made by the authors during field trips to Cuba in 2002 and 2004, and focuses on describing the characteristics of urban agriculture.

Since the introduction of urban agriculture in Cuba in the 1990s, a number of distinct categories have been defined, determined by size, location, users and yield. Table 17.1 is based on the situation in Havana but is typical for the country as a whole.

Given that the environmental case for urban agriculture relies on organic and local food production (see Chapter 3), we have concentrated on investigating high yield urban gardens, referred to as 'organpopnicos' in this text.

Organopnicos provide the highest yield of all forms of urban agriculture, and as such contribute the most to the horizontal intensification of a site. Their location within the urban fabric and the practice of selling crops from the farm gate, adds to their convenience for consumers and explains their presence in small rural towns.

In order to build up a picture of the qualities of urban agriculture, its spatial characteristics have been investigated at three different scales:

- the city
- the urban agriculture site
- the human.

To understand the implications for planning a city, which includes urban agriculture, we investigated the infrastructure required to support it. We also wanted to find out if organic agriculture was practised in Cuba.

Although the examples investigated in Cuba are influenced by local conditions, internationally relevant conclusions can be made. Local climate, topography and soil type all affect crop type, yield and the plot size required for a given return. Land ownership, neighbourhood and municipal boundaries all determine legal or social boundaries which may further impact on the size and location of urban agriculture sites, displacing sometimes 'randomly' the desired location of urban agriculture fields or their boundaries. These particular circumstances of location will interfere with general strategies for the design of Continuous Productive Urban Landscapes. Rather than being understood as negative realities, they are the factors which provide the local variations that lend interest to Continuous Productive Urban Landscapes.

THE SPATIAL CHARACTERISTICS OF CUBAN URBAN AGRICULTURE

At the scale of a city one can observe the relative distribution of urban agriculture sites in relation to each other and to other urban land use. At the scale of a single urban agriculture site, layout, form and materiality can be observed, and finally, at the human scale, the edges across which interactions occur, between citizen, cultivator and cultivated landscape, can be observed.

One of the issues we wanted to establish was how big urban agriculture plots need to be to provide full-time employment and be economically viable.

Table 17.1

	Size	Location	Farmers	Use of crops	Yield
State farms for producers consumption 'Autoconsumos Estatales'	1 hectare or more	Peri-urban	Voluntary cultivation by workers	Feed state workers, support day-care centres, homes for elderly, and facilities for new born babies, surplus sold to workers	1996: 0.34 kg/m².yr 2000: 0.6 kg/m².yr
Community gardens (plots) 'Huertos Populares (Parcela)'	Less than 1000 m²	Urban or peri-urban, vacant lots, unexploited area within educational or health facilities. State owned or private	One person or family	To supply cultivator or family	1996: 1–2 kg/m².yr 2000: 8–12 kg/m².yr
Community gardens (intensive cultivation garden) 'Huertos Populares (Huerto Intensivo)'	Typically between 1000 m² and 3000 m²	Urban or peri-urban, state owned or private land	One person or family, several families or co-operative	Feed producers and for trade	1996: 1–2 kg/m².yr 2000: 8–12 kg/m².yr
Urban community garden 'Organopónicos Populares'	Typically between 2000 m² and 5000 m²	Vacant urban sites, not suitable for direct agriculture use, require imported soil and containers	Groups of individuals formed into a collective. Institutional technical support and advice	Produce for trade and small-scale consumption by producers	1996: 3 kg/m².yr 2000: 20 kg/m². yr

Table 17.1 continued

	Size	Location	Farmers	Use of crops	Yield
High yield urban gardens *Organopónicos de Alto Rendimiento'*	Typically over 10 000 m²	Government allotted vacant urban sites, not suitable for direct agriculture use, soil and containers for growing brought in	Commercially viable work centres or co-operatives	Produce for sale to the population and tourist sector	1994: 12 kg/m².yr 2000: 25 kg/m².yr

Source: Caridad Cruz and Sanchez Medina (2003)

In addition to this we were interested in the precise layouts adopted to facilitate efficient growing and the patterns of crop rotation to judge how seasonality is reflected on the ground.

At a smaller scale we wanted to observe the interface between the urban agriculture plots and the city's inhabitants. We were interested in finding out how the city and its population interact with these plots.

Our studies began in Havana, and we then moved to Cienfuegos, a provincial city south of Havana, which has been referred to as the capital of urban agriculture in Cuba (Socorro Castro, 2001) and finally we visited Rodas, a small rural town near Cienfuegos.

URBAN AGRICULTURE AT THE CITY SCALE

Urban agriculture sites tend to be found on the urban fringe and in the city centre adjacent to major through roads, see Figure 17.1. Local conditions alter the relative distribution of urban agriculture. As density of inhabitation increases, for example as found in Havana, so city centre agriculture diminishes, while the relatively low density of Cienfuegos means that urban agriculture sites are available close to the centre of the city.

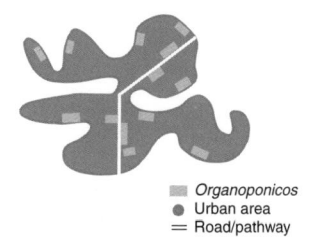

Cuban cities can be characterised as having enfolded edges. Urban agriculture is found predominantly alongside main thoroughfares and on the urban fringe

Figure 17.1

Havana has a European urban character with a compact core and a less dense, at times dispersed, edge. Its historic quarter is dense and most buildings do not exceed four or five stories, squares mark urban centres and the sea provides an edge. The pattern is typical for a city that has developed over time and a number of different planning strategies are evident in its layout. In common with many

PLANNING FOR CPULs: INTERNATIONAL EXPERIENCE

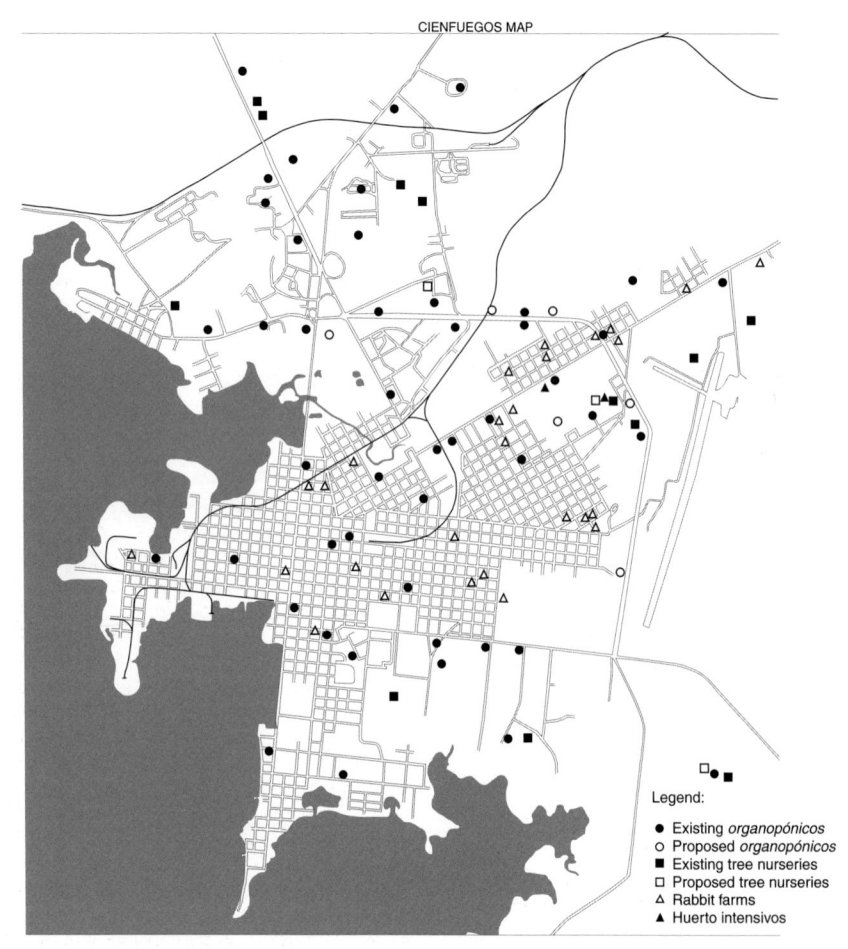

Figure 17.2

Figure 17.2 *Urban Agriculture sites in Cienfuegos.*

other cities, derelict plots of land in the city centre, often no larger than the footprint of a single building, have been converted into small-scale urban agriculture fields. These small plots are dispersed within the city's historic fabric, and can be viewed as a form of agitprop advertising the city's new means of food production, not unlike the programmes during the Second World War to convert public spaces in London to productive landscapes.

Moving out from Havana's centre, as the city de-compacts towards its edges, larger urban agriculture sites are found, as if enfolded into the city, often adjacent to industry or new residential developments. The largest urban agriculture fields, typical examples of peri-urban agriculture, are located on the urban periphery adjacent to main access routes into the city.

Cienfuegos provides a contrast to Havana. Whereas Havana has grown incrementally, Cienfuegos is a planned city, designed by French urbanists following a strict, but generous, grid pattern. It is like a provincial version of nineteenth century Barcelona, less dense, with one- or two-storey buildings, and complete with its own version of the Ramblas. The arrangement of urban agriculture in Cienfuegos reflects its layout. Many of its plots remain undeveloped and so the grid of streets lies over the land as a matrix into which urban agriculture sites can be inserted. Relatively low urban densities and the availability of land result in an even distribution of the population and urban agriculture sites.

Notwithstanding the availability of land in Cienfuegos, strict criteria are applied before a piece of land can be designated as suitable for urban agriculture. During an interview with Héctor R. López Cabeza, a member of the Provincial Group for Urban Agriculture in Cienfuegos, the following selection criteria were identified:

1. Sites must be empty.
2. Sites must not have been used for any other purpose.
3. Sites should fit into a long-term plan being developed by the municipality which integrates agriculture with physical, health and hydraulic resource (water) planning.
4. Water and electricity must be available on the site and the requirements for cultivation must not adversely effect the rest of the city.
5. It is very important that sites are close to consumers so that transport requirements are minimised.
6. If any of the above criteria are not met, then sites will not be designated for urban agriculture use.

Planning guidance for urban agriculture is decided locally and in cities other than Cienfuegos it has been located on sites which have had former uses. In Rodas, for example, the La Terminal urban agriculture site is found on an old bus terminus. In these cases, where soil may be contaminated, raised beds are used and imported soil, supplemented by compost, is used for growing food crops. The reason for requiring land with no previous use, in Cienfuegos, is unclear. It may stem from a concern about contaminated land, or it may have been introduced as a way of avoiding the need to clear land of buildings and other constructions prior to establishing agriculture. The impact of these criteria is significant – in Cienfuegos 1500 hectares of open space were identified as potential urban agriculture sites, but only 50 hectares met all of the selection criteria. This small proportion of land, classified as suitable for urban agriculture, in part reflects current difficulties in Cuba related to a lack of resources; for example, difficulties extending the existing water and electricity supply network, the cost of importing soil where there is contaminated land, and difficulties transporting goods over relatively short distances. Similar difficulties are likely to be found in many cities across the world. Their impact and significance will vary with the resources available to deal with them and

PLANNING FOR CPULs: INTERNATIONAL EXPERIENCE

Figure 17.3

Figure 17.3 *Urban Agriculture sites in Rodas.*

the degree to which urban agriculture is being integrated into an existing city or into a planned urban expansion onto green field sites.

A major research programme investigating the practice and development of urban agriculture in Havana took place between 1995 and 1998 (Caridad Cruz and Sanchez Medina, 2003). Part of the research programme investigated land management and criteria for inserting urban agriculture into Havana. The emphasis of this research focused on articulating the benefits of urban agriculture as a legitimate urban land use. While not underplaying the technical and legal constraints on selecting sites for urban agriculture, it emphasises the need to develop coherent design strategies for its integration into the city.

Rodas has been actively perusing urban agriculture since the start of the 'special period' in Cuba and, apart from producing fruit and vegetables, cut flowers are also grown within the town. In 2002, the town was starting to plant communal orchards flanking the main access road and a river, which runs through Rodas. This is one of the first examples we are aware of, where urban agriculture is consciously being developed as a part a wider landscape strategy. This can be compared to Havana's plans in the 1960s for a green belt, and offers a micro version of a productive landscape, providing environmental and quality of life benefits.

THE URBAN AGRICULTURE SITE

The size of urban agriculture sites in Cuba relates both to their location in the city and the type of urban agriculture practised (see Table 17.1 and Chapter 16).

Organopónicos populares are the most visible type of urban agriculture, although not the largest by total area or output. They are high yield urban market gardens, run commercially by their operators, selling vegetables from the 'farm gate'. In areas where vegetable production has been established, there are often floral organoponicos, producing cut flowers and garden plants. The size of organoponicos vary depending upon site availability and the number of people farming them. Cultivation is entirely manual, often using self-made tools. The smallest organoponicos have a cultivated area in the order of 500 m^2, which corresponds to the maximum area one person can

Main road to city centre

Productive urban landscapes as bridging elements

Figure 17.4

work – for comparison, a standard allotment site in England has an area of 250 m². The La Terminal organoponico in Rodas is fairly typical, with three people cultivating 1200 m² of raised beds. Significantly larger organoponicos do exist, for example the 4 Caminos site, which has a planted area of 3400 m² and a correspondingly larger workforce.

In organoponicos raised beds are used to isolate crops from soil which might be contaminated, and to contain the volume of soil to which compost is applied. The raised beds also provide healthier working conditions as the amount of bending required is reduced. A surprising amount of space is required for access paths and ancillary buildings. It is not unusual for a growing area of about 1000 m² to require a site of 3000 m². Peripheral areas of sites, initially unplanted, are often cultivated over time, thus increasing the efficiency of plots.

'Autoconsumos estatales' are similar to organoponicos, but are sited within the grounds of factories or other institutions and supply their local food needs. 'Huertos' or 'Parceleros' form another category and are small-scale plots usually cultivated by a single family; they are equivalent to allotments in England, although each cultivator's plot is often larger.

In reality, the differences between these categories are blurred. Nevertheless they do provide some insight in to the different forms of urban agriculture.

In 2000, in Havana alone there were more than 22 000 producers involved in cultivating 8 778 hectares of land. Table 17.1 shows the variation in yield between more traditional state farms and high yield urban gardens. Productivity for all types of urban agriculture has increased dramatically during Cuba's special period, for example, a doubling of productivity between 1994 and 2000, from 12 kg/m²kg to 25 kg/m²kg for high yield urban gardens.

Nursery houses, producing fruit trees and seeds are found either within organoponicos or on separate sites. Their output supplies the local population with plants for gardens and urban agriculture sites.

Figure 17.4 *Organoponico La Calzada in Cienfuegos.*

CUBA: LABORATORY FOR URBAN AGRICULTURE

Riverbank and recreation

PLANNING FOR CPULs: INTERNATIONAL EXPERIENCE

Urban agriculture fields provide an educational resource for adjacent schools.

Figure 17.5

Figure 17.5 *Organoponico Gastronomica Playa in Havana.*

CUBA: LABORATORY FOR URBAN AGRICULTURE

Translucent enclosures suggesting the kinds of spaces which could be developed within CPULs.

Figure 17.6

Observations and speculations

In a few words we can tell you how to build an organoponico.

Divide a site up into alternate strips, 65 cm wide for paths, 120 cm wide for raised beds. Excavate the beds to a depth of 30 cm below ground, fill with stones for drainage and build retaining walls to a height of 20 cm above ground. Fill beds with soil and repeat until the site is full.

On the face of it, one organoponicos is much like another. But when one looks, it can be seen that this is not so. The particularities of each site interfere with and enrich the apparent monotony by distorting and varying it.

The urban agriculture atlas which follows, presents surveys of ten sites measured by the author during February 2002. The text accompanying the survey drawings describe conditions as found, and characteristics which could inform future design strategies. The site surveys are followed by a record of the materials used to construct planting beds for organoponicos. In the ten sites surveyed, a mix of twenty-four materials was found, defining the edges of planted beds. In all cases these were recycled materials, some placed loosely for easy reassembling, some placed permanently. This variety of materials and the changing crop patterns over time contribute to the visual dynamics of urban agriculture and its role as a register of seasonal change.

The edges of urban agriculture sites form boundaries between the public realm of the street and the private realm of market gardens. They also offer the potential to become occupied. Security is an important issue for cultivators and a fence of some sort always surrounds urban agriculture sites in Cuba. Although these fences provide a physical barrier along boundaries, they do not prevent passers-by from looking at and sensing the urban agriculture sites. Different sites and different boundaries presented different characteristics, often determined by the ad hoc appropriation of edges for different uses, for example, the use of a fence for drying clothes. Observations in Cuba show how boundaries can offer new places for occupation and new experiences within the city.

Figure 17.6 *Plant nursery in Cienfuegos.*

Spaces capable of accommodating temporary occupation (a wedding) or transitory occupation (sport)

A path or road often bound one or more sides of an urban agriculture site. Networks for circulation and distribution are established, in some cases a very particular quality is generated by the simplest of elements or activities. A wall can mark a route and boundary, while its shadow creates a three-dimensional comfort zone, a space providing thermal delight, encouraging a different pace and making a space to talk while viewing an ever changing landscape.

By looking out over productive urban landscapes from a window or terrace, the observer gains ownership of their territory; it is claimed by seeing. Environments become comprehensible, an urban nature is defined, personal space is expanded, from your window, and from your neighbour's window as well!

PLANNING FOR CPULs: INTERNATIONAL EXPERIENCE

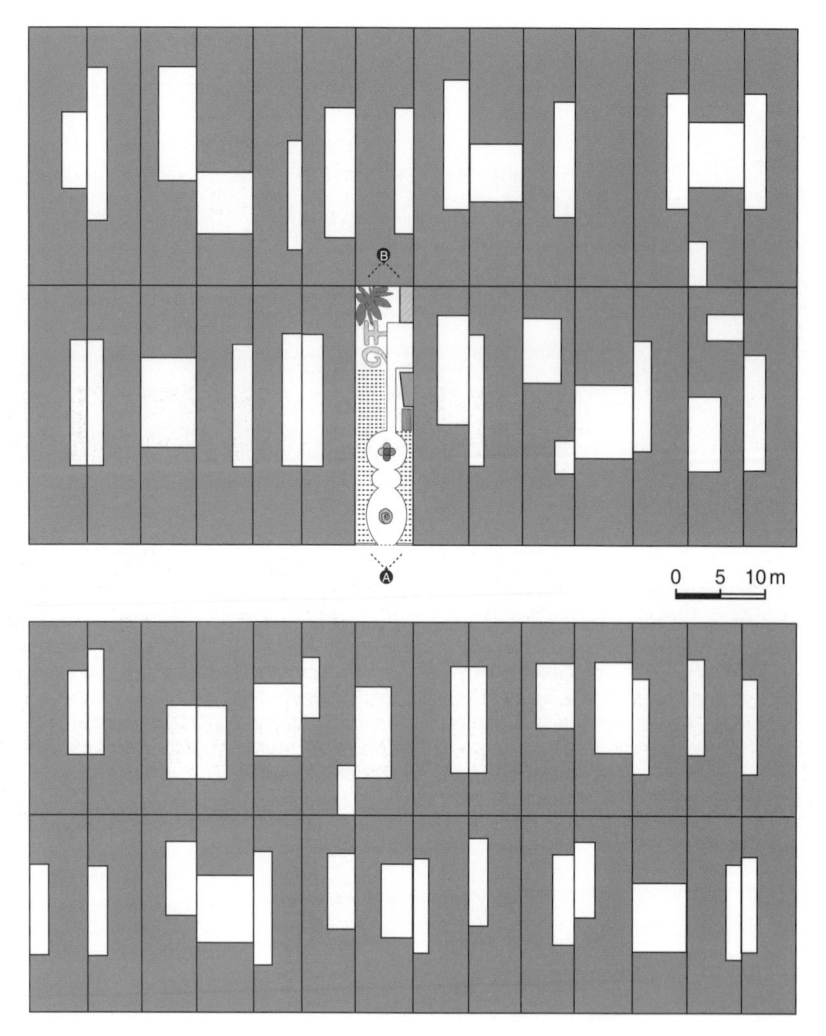

Figure 17.7

CUBA: LABORATORY FOR URBAN AGRICULTURE

CHARACTERISTICS	HUERTO COMUNITARIO SAN MIGUEL, BELASCORIN Y GERVASIO. CENTRO HABANA
Micro void Enclosure Ground Walls Sky A place to sit The smell of earth	A community garden, established by a group of local women. This is one of the few examples of a permaculture garden in Cuba. The geometry is non-linear, leading occupants from the entrance towards seating spaces.

CROPS	MATERIALS
avocado pears, furrows for beans, pumpkins, banana tree, aloe, peas, fennel, chillies, herbs	stone, earth, gravel, car tyres, tiled path

Figure 17.8

PLANNING FOR CPULs: INTERNATIONAL EXPERIENCE

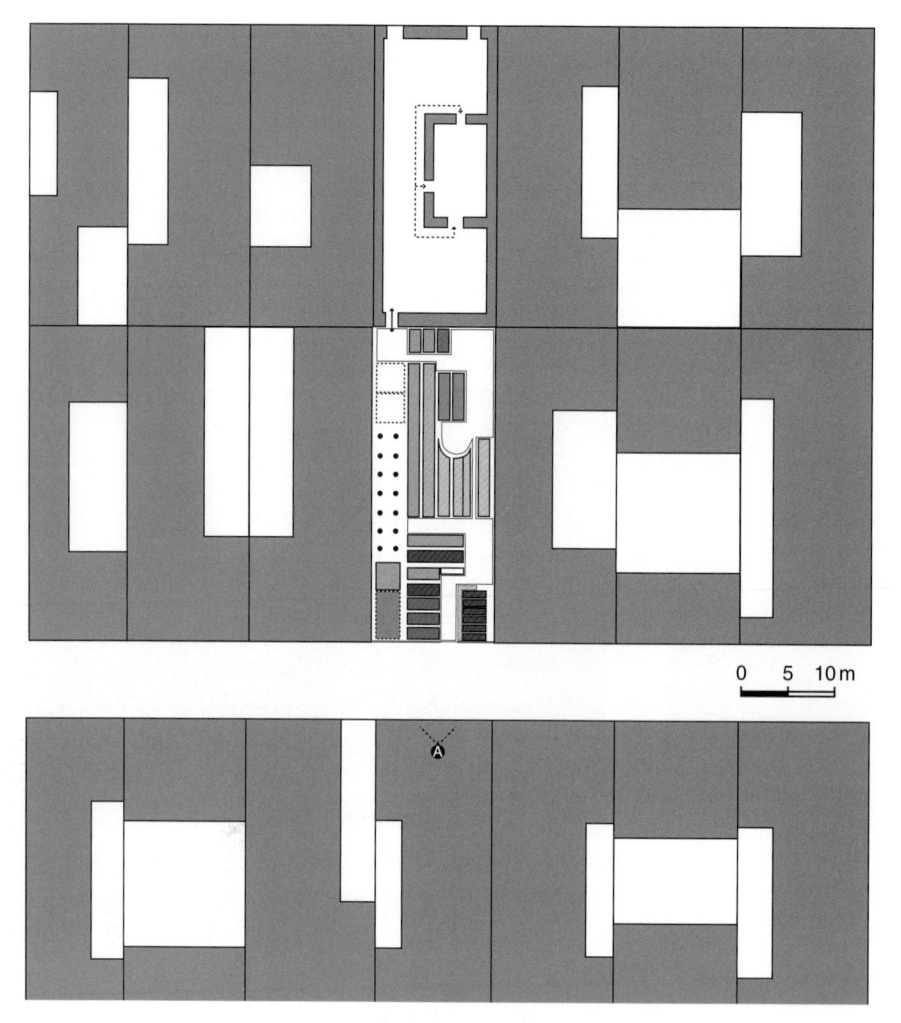

Figure 17.9

CUBA: LABORATORY FOR URBAN AGRICULTURE

CHARACTERISTICS

HUERTOS INTENSIVO HABANA, MERCED Y PAULA. HABANA VIEJA

Linear micro garden
Outdoor class room
Debating chamber
Valley section
Marking space with shade
A shared visual facility
Terraces
Windows
Balconies

This site, in the historic quarter of Havana is managed by Alberto de la Paz. It provides food and functions as an educational facility for local school children, who visit and work on the crops. A tree shades an outdoor meeting space.

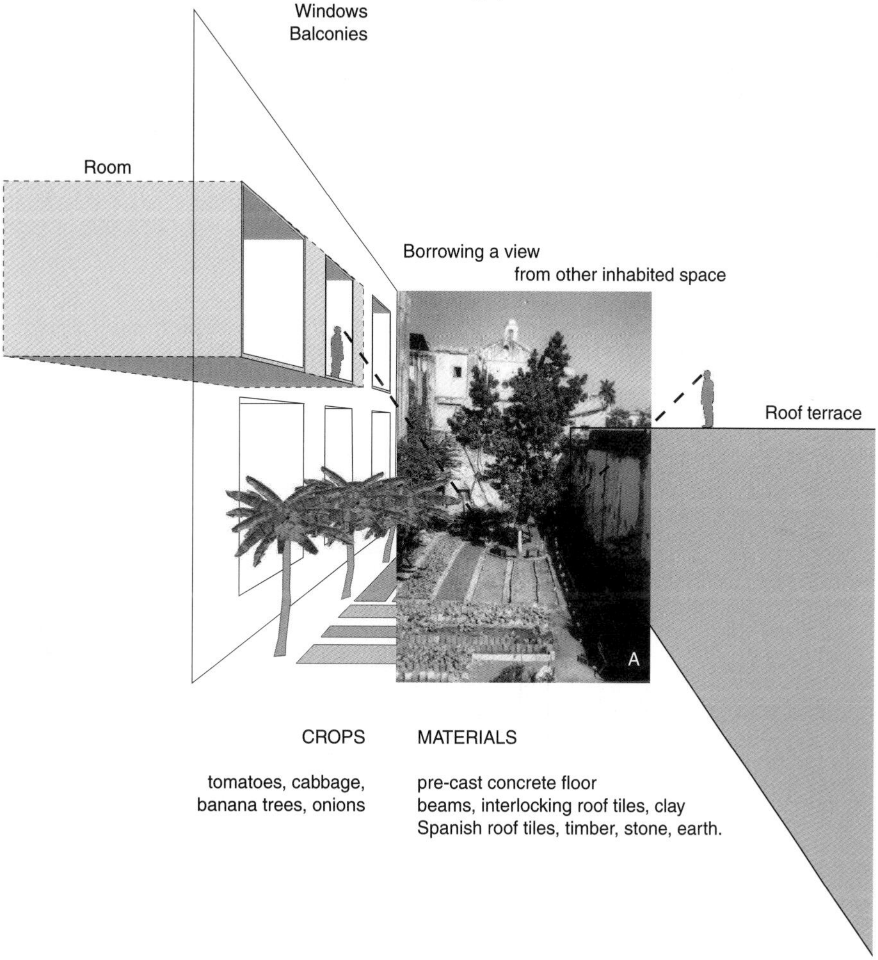

Room

Borrowing a view from other inhabited space

Roof terrace

CROPS

tomatoes, cabbage, banana trees, onions

MATERIALS

pre-cast concrete floor beams, interlocking roof tiles, clay Spanish roof tiles, timber, stone, earth.

Figure 17.10

PLANNING FOR CPULs: INTERNATIONAL EXPERIENCE

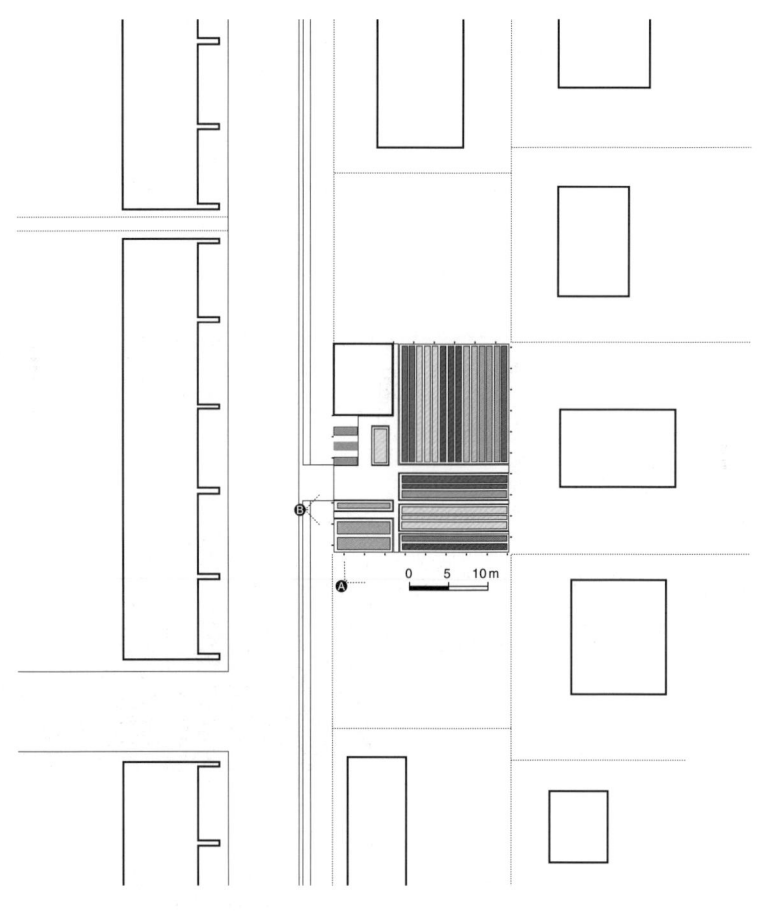

Figure 17.11

CUBA: LABORATORY FOR URBAN AGRICULTURE

CHARACTERISTICS

Productive landscape patch
Suburban plot as market garden
Alongside a main access road
Thickened boundary
Layered edge

ORGANOPONICO PUEBLO GRIFO VIEJO
CIENFUEGOS

A site for home working, this small organoponico is located within the curtilage of a suburban plot. It is small enough for one person to be able to cultivate it. At this scale it does not challenge, or react to the characteristic conditions of sprawl, found in adjoining plots.

CROPS

lettuce, tomatoes,
onions, maize

MATERIALS

recycled precast concrete beams,
40 cm by 40 cm by 5m long, earth

Figure 17.12

PLANNING FOR CPULs: INTERNATIONAL EXPERIENCE

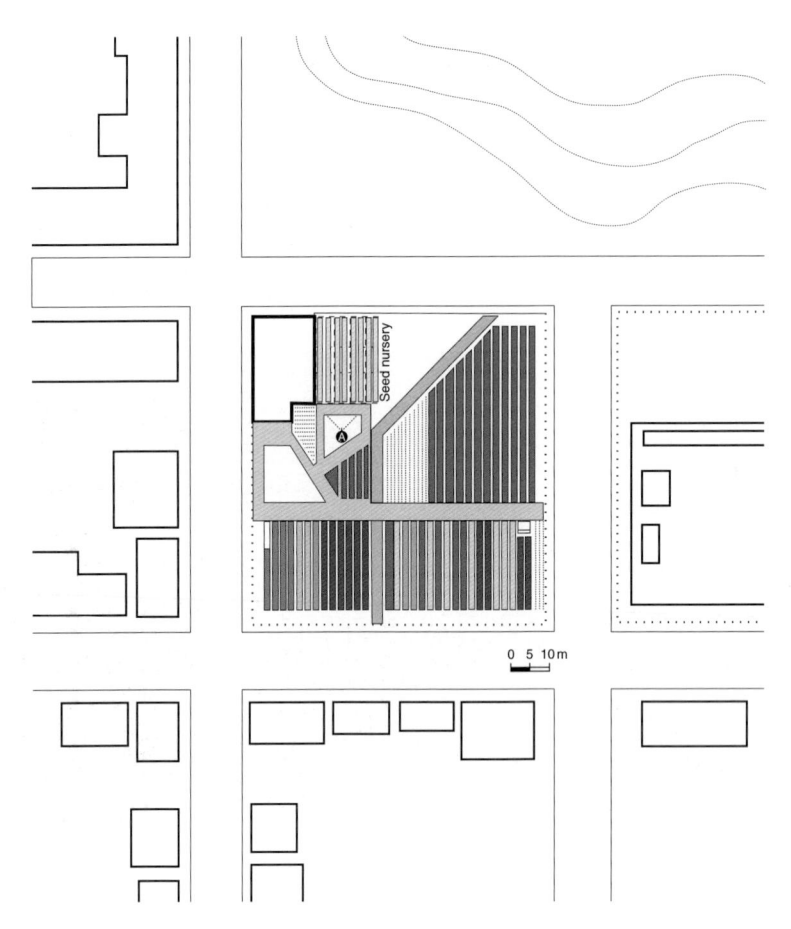

Figure 17.13

CUBA: LABORATORY FOR URBAN AGRICULTURE

CHARACTERISTICS	ORGANOPONICO LA TERMINAL RODAS
Brown field site Urbanised field Cultivated city block Horizontal Corner shop Health advice Translucent enclosure	The terminal occupies the site of a previous bus terminal. Located within a city block, close to the edge of a small rural town, called Rodas. Placed like a stepping stone in a steam, it will eventually form the first of a series of such fields leading to community orchards flanking two rivers that define Rodas. An enigmatic space is created by a translucent enclosure containing a seed nursery.

CROPS	MATERIALS
lettuce, onions, radish, beans, tomatoes, carrots, medicinal crops, hibiscus	stones, wood, concrete blocks, concrete posts, zeolite, earth, soil, slender steel framework, railway tracks for columns, reinforcing rods supporting shadow mesh.

Figure 17.14

PLANNING FOR CPULs: INTERNATIONAL EXPERIENCE

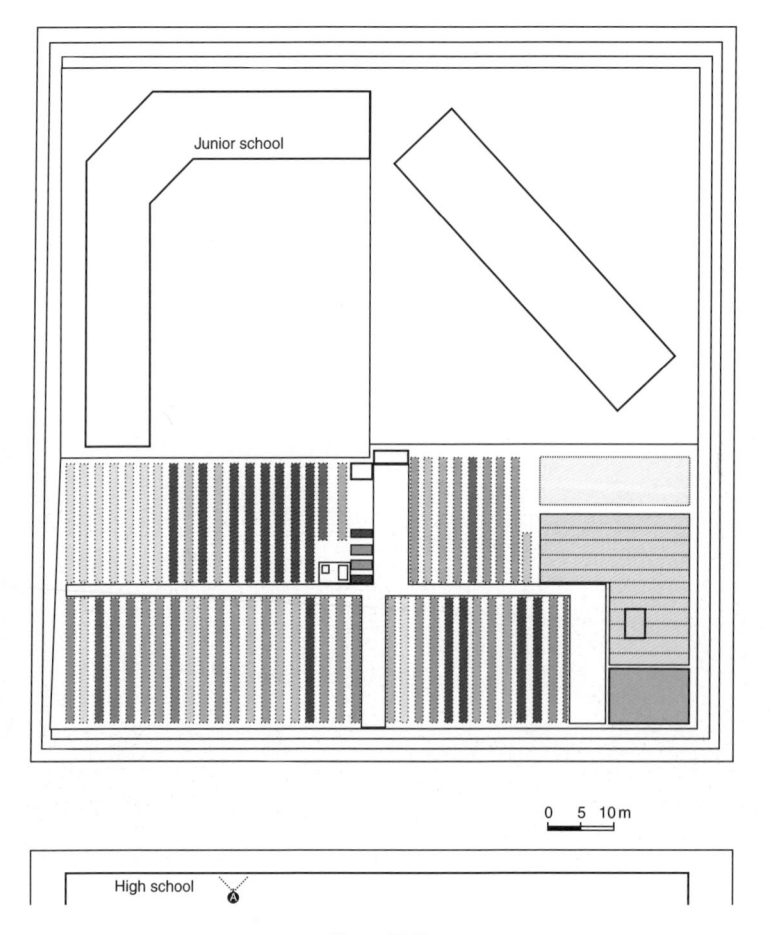

Figure 17.15

CUBA: LABORATORY FOR URBAN AGRICULTURE

CHARACTERISTICS

ORGANOPONICO GASTRONOMICA PLAYA
MUNICIPIO PLAYA. HAVANA

Plateau
Shared resource
A space between
Boundary and edge
Urban carpet
Educational resource
Embankment to pavement

Playa organoponico serves a predominantly residential area. It is defined by being constructed on an artificial plateau sited on sloping ground. Pedestrians walking around its perimeter experience a constantly changing relationship with the urban agriculture surface. Sited between a single storey junior school and a multi-storey high school, it is seen as a carpet from one and an edge to the other.

CROPS

tomatoes, lettuce, parsley, cabbage, chives, leeks, onions, rocket, peppers, aubergines, beans, cauliflower, pumpkin

MATERIALS

stone, concrete test cores, concrete rubble, corrugated cement sheet, chain link fence, earth

Figure 17.16

PLANNING FOR CPULs: INTERNATIONAL EXPERIENCE

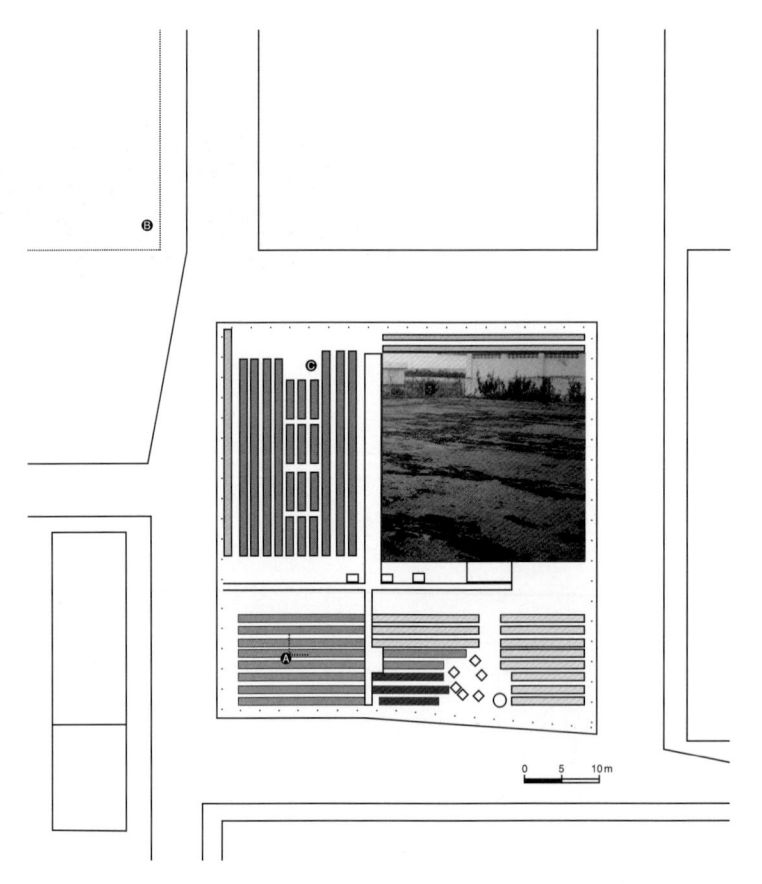

Figure 17.17

CUBA: LABORATORY FOR URBAN AGRICULTURE

CHARACTERISTICS

Platforms for sport, performance and cultivation.
Play between horizontal and sloping ground.
Made ground and found ground
Wide horizons
Views across territories
A visual invitation to occupy
Patchy landscapes

AUTOCONSUMO
UNIVERSIDAD DE CIENFUEGOS

This market garden supplements the fruit and vegetable requirements of the University of Cienfuegos. It suggests how productive landscape can connect and create transitions between different functional uses. By encompassing non specific territories, occupation is invited. For example the reinforced concrete top to a water tank provides circa 800 sq. m of performance space.

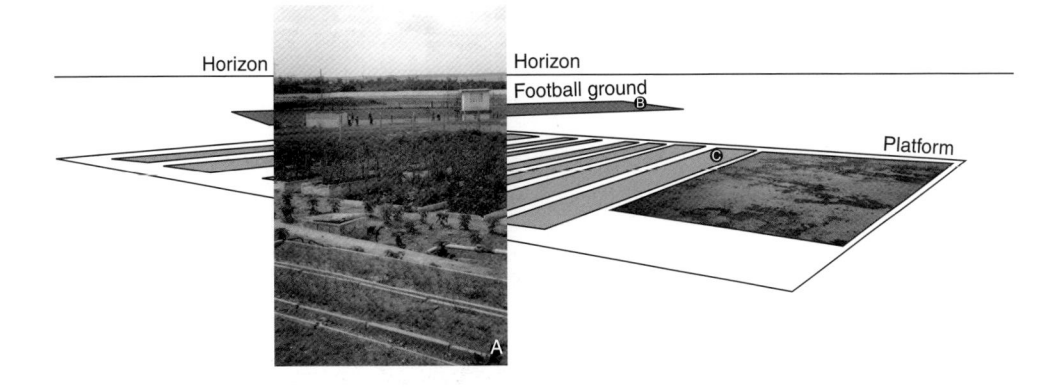

CROPS

peppers, okra beans, lettuce,
tomatoes, chives, aubergines

MATERIALS

precast concrete slabs held
in place with metal hoops, some recycled
stool legs, tar macadam, earth, chain-linkfence,
precast concrete posts

Figure 17.18

PLANNING FOR CPULs: INTERNATIONAL EXPERIENCE

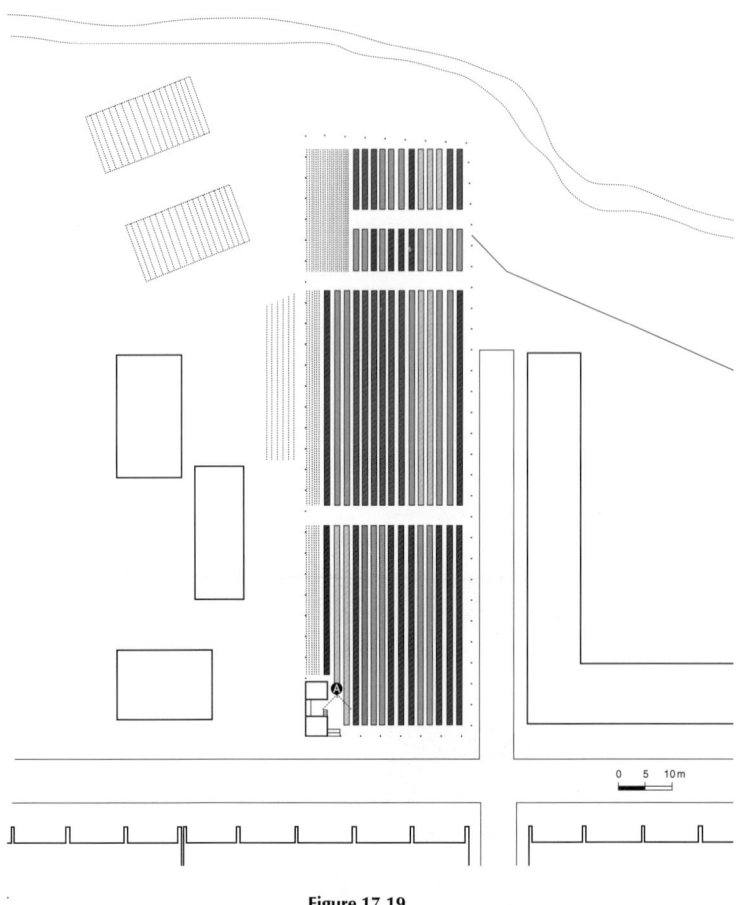

Figure 17.19

CUBA: LABORATORY FOR URBAN AGRICULTURE

CHARACTERISTICS

Landscape carpet
Intensified view
Changing urban surface
Improving accessibility
Linking
Marking a new urban axis
Market stall as threshold

ORGANOPONICO LA CALZADA
CIENFUEGOS

An organoponico with potential for introducing a coherent landscape strategy, connecting major circulation routes to a leisure landscape alongside a river which runs through the city.
The introduction of public routes through or alongside urban agriculture fields would provide a new dimension to the city and be an example of ecological intensification.

Same level view   different to

Higher level view

CROPS

lettuce, beans, onions, fallow land,
seed beds, tomatoes, coriander.

MATERIALS

rendered concrete blocks,
earth, soil.

Figure 17.20

PLANNING FOR CPULs: INTERNATIONAL EXPERIENCE

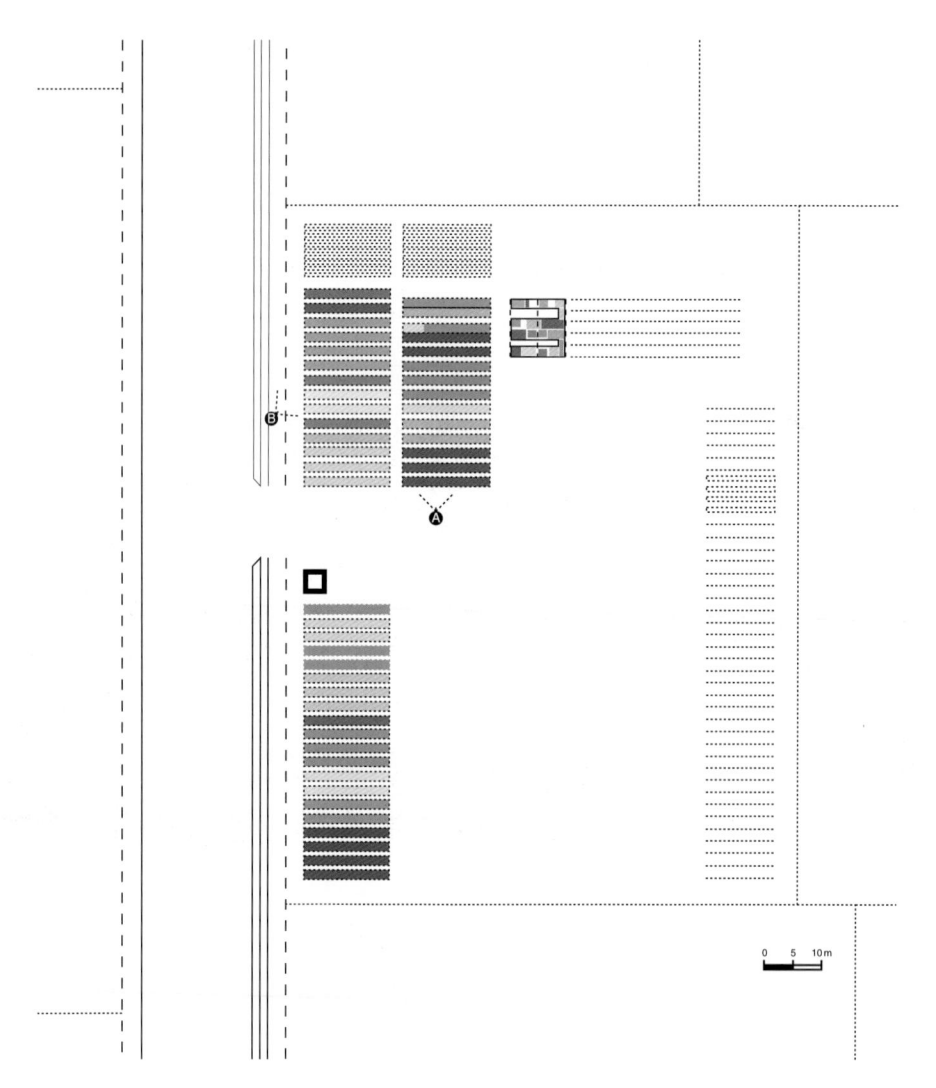

Figure 17.21

CUBA: LABORATORY FOR URBAN AGRICULTURE

CHARACTERISTICS

Incremental occupation
Bounding a horizontal site
Colour and scent
Indeterminate centre
Fairground
Market
A place for wedding receptions

ORGANOPONICO FLORAL RODAS

This site, located before one enters Rodas, supplies the town with cut flowers and garden plants. It has characteristics similar to fruit and vegetable organoponicos. The importance given to the provision of cut flowers demonstrates how an attitude to living can be embodied in productive landscapes. A loosely structured arrangement of beds define an uncultivated central area which invites transitory occupation.

CROPS

irises, poppies, daisies,
hibiscus, fruit trees

COLOURS

red, orange, green, pale red, purple,
white, green and white variegated,
dark green, lime green, grey,
grey green, reddish green

MATERIALS

stone, earth, timber posts,
woven plastic sheeting, rushes

Figure 17.22

PLANNING FOR CPULs: INTERNATIONAL EXPERIENCE

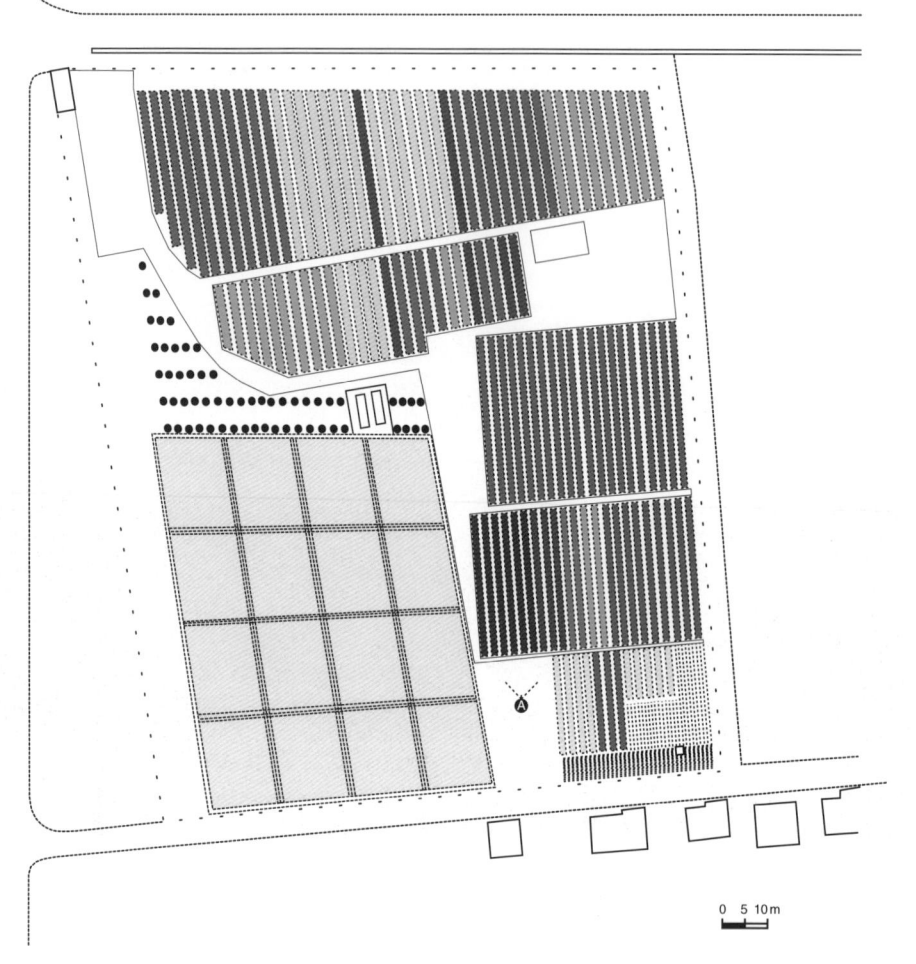

Figure 17.23

CUBA: LABORATORY FOR URBAN AGRICULTURE

CHARACTERISTICS

ORGANOPONICO CUATRO CAMINOS CIENFUEGOS

Extensive
Peri urban agriculture
Articulating the boundary between built and cultivated environments
Marking pedestrian and cycle routes
Making space with shade
Market stall

A main road bounds this extensive site to one side with pedestrian and cycle paths on its other sides. A wall three meters high separates the organoponico from the University of Cienfuegos campus and defines two territories representing, an urban condition and a rural condition. The wall shades a cycle path.
The upper floors of adjacent buildings overlook the wall, thus making a visual connection between elevated interiors and the rural condition represented by the urban agriculture.
On the ground the wall isolates and defines the campus space.

CROPS

lettuce, tomatoes, beans, beetroot, potatoes, chives, okra, radishes, aubergines, garlic, maize, onions

MATERIALS

stone, concrete wall, chain link fence, earth

Figure 17.24

PLANNING FOR CPULs: INTERNATIONAL EXPERIENCE

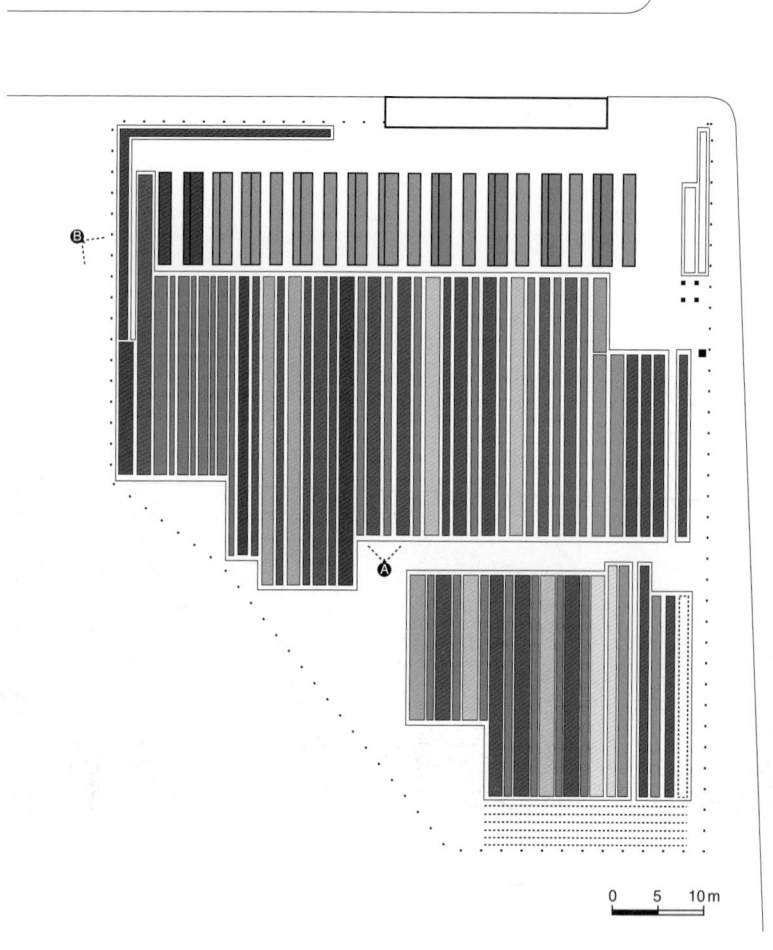

Figure 17.25

CUBA: LABORATORY FOR URBAN AGRICULTURE

CHARACTERISTICS	ORGANOPONICO PASTORITA CIENFUEGOS
Undulation Marking topography Making a new surface Viewing from above Bridging territories	This site is discovered from above, from private apartments or from shared terraces of an adjoining school. Recycled five meter long precast concrete beams enclose planting beds and articulate topographical changes, thus defining territories and edges. This landscape, currently a visual resource, could become a physical resource if overlaid with publicly accessible paths and spaces, increasing the variety of occupation.

CROPS	MATERIALS
aloes, mint, pumpkins, lettuce, spinach, parsley, beans, beetroot, tomatoes, carrots, coriander, citrus trees	precast concrete slabs, 400mm by 400mm precast concrete beams, concrete blocks, concrete fence posts, wire fence

Figure 17.26

PLANNING FOR CPULs: INTERNATIONAL EXPERIENCE

MATERIALS

Concrete test cores ❶
Organoponico Gastronomia. Playa. Havana

Precast concrete slabs/steel reinforced bars ❷
Pueblo Grifo Nuevo. Cienfuegos

Hollow concrete blocks/precast concrete slabs ❸
Cuatro Caminos. Cienfuegos

❹ Hollow terracota bricks
Universidad de Cienfuegos

❺ Precast concrete slabs/terracota bricks
Universidad de Cienfuegos

❻ Insitu concrete walls
La Calzada. Cienfuegos

Figure 17.27

CUBA: LABORATORY FOR URBAN AGRICULTURE

MATERIALS

Precast concrete slabs/concrete blocks **7**
Pastonia. Cienfuegos

Precast concrete beams/insitu concrete corner **8**
Pueblo Grifo Viejo. Cienfuegos

Plastic 'grow' bags **9**
Floral Organoponico. Rodas

10 Rough cast concrete slabs
Pueblo Grifo Nuevo. Cienfuegos

11 Precast concrete slabs/steel stool legs
Universidad de Cienfuegos

12 Natural stones
Floral Organoponico. Rodas

Figure 17.28

PLANNING FOR CPULs: INTERNATIONAL EXPERIENCE

MATERIALS

Interlocking terracota roof tiles ⓭ ⓰ Rough cast concrete/insitu concrete
Huertos intensivos. Havana   La Terminal. Rodas
Stone and crushed concrete ⓮ ⓱ Precast concrete/insitu concrete
La Terminal. Rodas   La Terminal. Rodas
Precast concrete beam/concrete blocks ⓯ ⓲ Terracota roof tiles
Huertos intensivos. Havana   Huertos intensivos. Havana

Figure 17.29

CUBA: LABORATORY FOR URBAN AGRICULTURE

MATERIALS

Crushed concrete and brick (19)
Organoponico Gastronomia. Playa. Havana

Rubber tyres (20)
Huerto Comunitario. Havana

Lime washed stones (21)
Cuatro Caminos. Cienfuegos

(22) Crushed concrete
Huerto Comunitario. Havana

(23) Concrete test cores
Organoponico Gastronomica. Playa. Havana

(24) Precast concrete columns
Pastorita. Cienfuegos

Figure 17.30

PLANNING FOR CPULs: INTERNATIONAL EXPERIENCE

EDGE DEVELOPMENT
THICK EDGE

Transitory occupation changes the way in which an edge is perceived. A fence hung with washing to dry demonstrates how the use of an edge changes the way a passer-by views the urban agriculture field beyond. Someone looking towards the field will focus alternately on the materials in the foreground and the background. In this situation one reads an adjacent surface, the fabric hung out to dry first and later the earth and crops beyond. As the fabrics change, so new visual relationships are set up between foreground and background, a constantly changing dialogue between the vertically hung cloth and the horizontal ground.

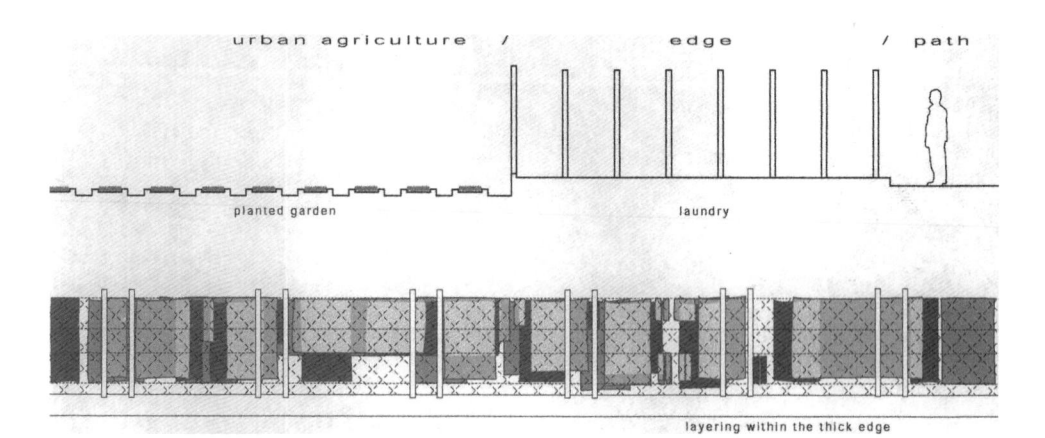

The fence of the Organoponico La Calzada, in Cienfuegos is used to dry laundry.

THICK EDGE

Imagine the effect of several rows of washing along the edge of an urban agriculture site; this is an example of a thick edge. Thick edges accommodate different types of inhabitation, separating private and public space and in so doing add to the urban experience.
Thick edges may accommodate a variety of occupations, for example, a surface for play, a garden for leisure or platforms for picnics.

EDGE DEVELOPMENT PLAN
(STAGE ONE)

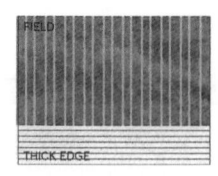

Figure 17.31

CUBA: LABORATORY FOR URBAN AGRICULTURE

EDGE DEVELOPMENT
DUAL NATURE EDGE

Windows in buildings on the edge of urban agriculture sites offer views with different characteristics. The view towards urban agriculture presents an image, which is rural in character. Windows facing other directions present an image urban in character. Thus, within a given building different conditions may be experienced, setting up different moods in different spaces.

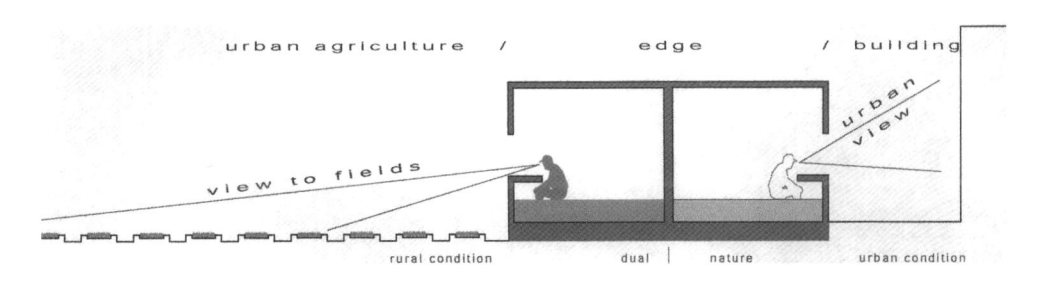

Alternating views from a staircase, to one side views of the Pueblo Grifo Nuevo organoponico in Cienfuegos and to the other side views of adjacent buildings.

DUAL NATURE

This transition from a rural or natural condition to an urban condition, which occurs in plan, is not unlike the articulation of space made by the classical division of a building vertically into base, piano nobile and attic.

EDGE DEVELOPMENT PLAN
(STAGE TWO)

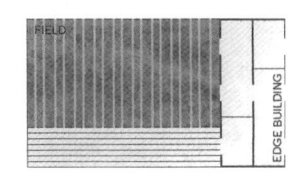

Figure 17.32

PLANNING FOR CPULs: INTERNATIONAL EXPERIENCE

EDGE DEVELOPMENT
THIN EDGE

Here a minor intervention, a wall 3 metres high, introduces a series of significant spatial divisions. The location and dimension of the wall simultaneously separates and links the organoponico to adjacent buildings. At ground level, the space between the wall and buildings is perceived as an urban area, like an urban square, or plaza, bounded and contained by walls and buildings. This space has an urban quality and intensity. On the other side of the wall, a person is shielded from the adjoining buildings and their view is concentrated towards the market garden. The wall intensifies the contrast between an urban and rural condition.
Pedestrians moving from one side of the wall to the other experience this contrast.

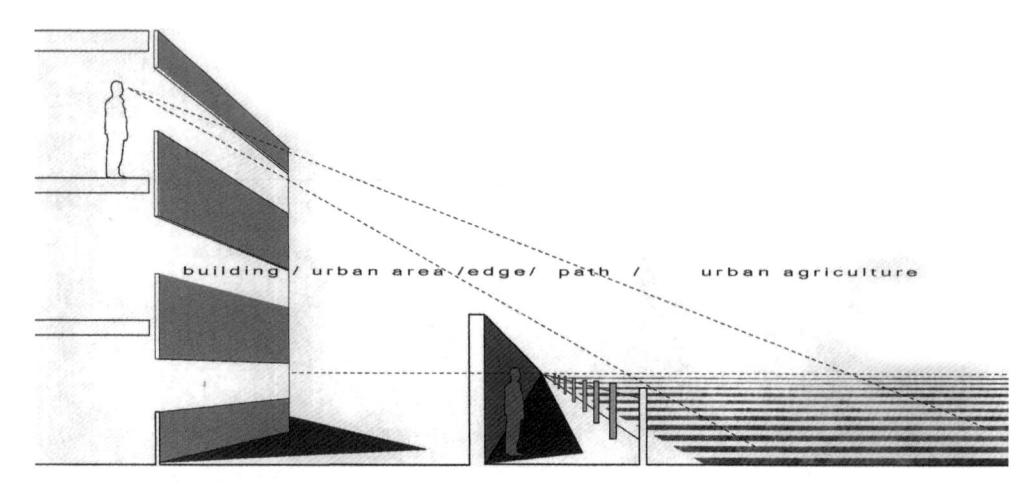

A cycle path shaded by a wall runs alongside the Cuatro Caminos organoponico in Cienfuegos.

building / urban area /edge/ path / urban agriculture

THIN EDGE

The change in experience can be understood as a marker, recording the change between the working day and domestic or relaxing time. The daily change between rural and urban condition is another example of urban intensification, introduced by the location of productive landscape. It can be compared to the weekly contrasting environment experienced by people occupying country cottages during weekends and town houses during the week.

EDGE DEVELOPMENT PLAN
(STAGE THREE)

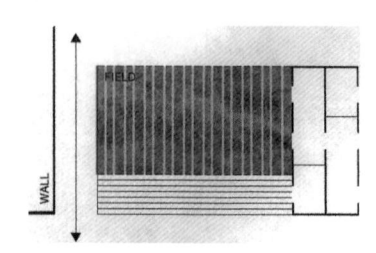

Figure 17.33

CUBA: LABORATORY FOR URBAN AGRICULTURE

EDGE DEVELOPMENT
TOPOGRAPHICAL EDGE

In several instances ground has been levelled to provide a horizontal surface for cultivation. Horizontal planes accentuate undulations in the surrounding landscape and change the way in which an observer sees the market garden. These changes between the level of ground on which a pedestrian walks and the adjacent market garden mean that at times the field may be in view and at times it may be hidden. A sense of expectation is provoked in passers-by.

Embankments between pavements and market gardens at the Pastorita organoponico, Cienfuegos and the Organoponico gastronomia Playa, Havana.

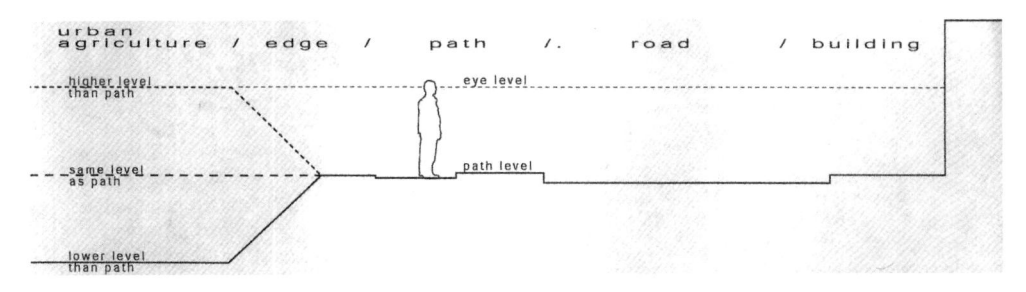

THIN LINE EDGE

Horizontal planes represent artificial topographies. By accentuating changes in the natural topography, a pedestrian's awareness of the specific character of the place is enhanced and measure is given to the landscape.

EDGE DEVELOPMENT PLAN
(STAGE FOUR)

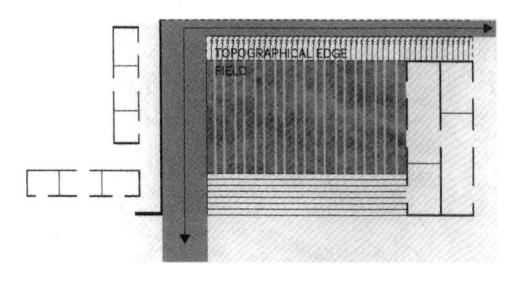

Figure 17.34

INFRASTRUCTURE AND URBAN AGRICULTURE

In this section three situations are examined. First, the infrastructure supporting urban agriculture at a municipal level (Rodas), second the role and observations of an independent foundation in Havana (Fundacion Antonio Núñez Jiménez De La Natualeza Y El Hombre) and finally an overview of the practice of urban agriculture in a particular neighbourhood (Consejo Popular Camilo Cienfuegos).

Rodas provides an example of how a small town has actively encouraged urban agriculture. The 'Programa Especial De Desarrollo de la Agricultura Urbana' supports a team of advisors helping farmers manage their sites and business. A meeting with this team, held in Rodas on the 16 February 2002, clarified the role of the municipality and urban agriculture advisors.

In 1994 urban agriculture was introduced to Rodas, the main town in Rodas municipality (population of 33 600). Initially only vegetables were produced, but the goal was expanded to provide fresh fruit and vegetables for all of the population. Currently all twenty-nine settlements in the municipality have productive urban agriculture sites.

Vets provide advice to urban farmers on animal husbandry and horticulturists advise on organic pest control, based on no use of chemicals. A programme also exists to supply urban farmers with organic fertiliser from two sugar mills, and cattle raising and chicken farming opportunities.

The office for urban agriculture will supply families with ten chickens and one rooster for egg production. In addition to food growing sites, there are floral organoponicos (for cut flowers), seed farms and nurseries. Communal orchards have previously been established, and fruit trees are also supplied for planting in private gardens and a small pineapple farm has been established to supply the tourist trade.

Urban agriculture sites are partnered with schools, medical centres and old people's homes and contribute to programmes promoting healthy eating. School clubs have been set up, with programmes developed in consultation with the pupils.

It is planned to concentrate vegetable production in organoponicos rather than intensive farms, as their yields are higher. Yields from intensive farms have been found to be 1 kg of vegetables per square metre per month, while organoponicos yield about 2.5 kg of vegetables per square metre per month. These yields which are high (see Chapter 20) are due to the intensive farming of the organoponicos and in part due to Cuba's climate which allows multi-cropping per year.

In 2001, the town of Rodas was self-sufficient in fruit and vegetable production. Output has increased from 350 grams of fruit and vegetables per inhabitant per day in 1994 (127 kg per year) to 1600 grams in 2000 (584 kg per year). The aim is to increase the output to 2000 grams per inhabitant per day (730 kg per year). This level of production compares with 115 kg per inhabitant per year in 2000 for Havana (Caridad Cruz and Sanchez Medina, 2003).

In ten years' time the province of Rodas aims to:

- be self-sufficient in fruit and vegetable production
- be able to export organic crops
- have trees in towns to help provide fresh air.

In Havana, state and municipal support for urban agriculture is complimented by the work of the Fundacion Antonio Núñez Jiménez De La Natualeza Y El Hombre. The foundation was established by the founder of the Cuban Academy of Sciences, and believes that only through culture can the relationship between humanity and nature be ameliorated. Thus, promoting urban ecology within a broad context of environmental and social sustainability, is one of its many activities and as part of this work, urban agriculture is supported (Caridad Cruz and Sanchez Medina, 2003). Permaculture and organic urban agriculture are vigorously promoted by working with local communities and supporting local initiatives by means of workshops and networking (see Chapter 22). The Foundation provides capacity building programmes for the urban agriculture community and since 1994 has promoted permaculture, which was introduced to Havana in 1993 by practitioners from New Zealand

and Australia. Six or seven small-scale projects were set up in Havana, followed by 43 permaculture farms, although not all of these farms have proved successful. In practice permaculture remains a fringe activity, which is not integrated into wider, state supported programmes.

A meeting in Havana on the 18 February 2002, with Roberto Sánchez Medina, who manages the Fundacion Antonio Núñez Jiménez De La Natualeza Y El Hombre's sustainable cities programme, provided insights into the role of urban agriculture in Havana.

As yet Havana is not self-sufficient in fruit and vegetable production. A potential 28 000 hectares are available for urban agriculture. Up to the year 2000, excluding state enterprises, about 8000 hectares were used for urban agriculture. Of this about 1500 hectares were located within the city of Havana, but most urban agriculture is located on the urban periphery or in the suburbs.

The Cuban government defines urban agriculture by the distance of fields from the city. The limits are set at 10 km for a typical city and 5 km for a small city. Havana is a special case and any food growing within the metropolitan area is considered urban agriculture.

There are four large parks within Havana but since the special period few resources have been available to manage these spaces. The largest of these, the Parque Metropolitano de la Habana, already contains some urban agriculture and there are plans for the future expansion of urban agriculture in these parks.

Since 1994, urban agriculture in Havana has had support from the department of agriculture and there are now 17 urban agriculture advice shops in the city. These support individuals who wish to establish urban agriculture sites. The state supplies land to individuals, or groups of individuals who sign a contract to produce particular crops, which may be sold on the open market. A rental for the land is paid to central government.

The improving economic situation has provided limited access to pesticides, but the law prohibits the use of most chemicals. A few very specific chemicals are permitted for use in a controlled manner.

It was noticeable that the people we spoke to in Havana were more cautious about predicting the future for urban agriculture than those in Rodas. They observed that changes in the economic order could alter its status and perceived importance. One of the major issues to face in the future will be the value and ownership of land. Currently this is not a problem, as sufficient state owned land exists.

The feeling was that if the benefits are well understood, then urban agriculture has a good chance of succeeding.

URBAN AGRICULTURE IN RESIDENTIAL AREAS

To better understand how urban agriculture is promoted in residential areas, the Fundacion Antonio Núñez Jiménez invited us to visit the Consejo Popular Camilo Cienfuegos, a large residential settlement on the eastern edge of Havana, built after the revolution, housing some 11 600 people. The settlement's people's council oversees the implementation of urban agriculture.

In February 2002, a total of approximately eight hectares of land were under cultivation, consisting of the 'El Pedregal Intensive Cultivation Garden' (2.5 hectares in which crops are sown directly in the ground, rather than in raised beds) and the El Paraiso Farmers Group, a group of 45 parcelas (allotments). While produce from a parcela is sometimes sold to the public, their primary purpose is to provide food for the individuals.

A further 4.3 hectares of land was cultivated by dispersed farmers, operating as individuals and not attached to the people's council.

The Consejo Popular Camilo Cienfuegos has an area of about six square kilometres, which is subdivided into nine sub areas, each of which elects a local representative to the people's council. Local representatives work directly with farmers and a single representative reports back to the agricultural administrators and planners at the provincial level. This system provides a robust mechanism allowing feedback between policy makers and individual urban farmers.

Participation in urban agriculture is voluntary and relies on individuals or groups making a request to area representatives for land to grow food on.

If an inhabitant wants land for a parcela, or a group wants land for an intensive farm, this is granted subject to conditions requiring the growing of food and prohibiting the construction of buildings or the cutting down of trees. A clear distinction exists between commercially run projects like the El Pedregal Intensive Cultivation Garden and non-commercial parcelas.

Parcelas tend to be cultivated by retired people and they are free (no rent); large ones may cover 2500 m^2 if a whole family uses it, but 200 m^2 is more typical.

In the case of Intensive Cultivation Gardens the Greater Havana Fresh Vegetable Company co-ordinates the sale and distribution of crops from a number of producers and collects a sales tax that the farmers are liable for.

In 2002 land remained available for urban agriculture. The ground around the Consejo Popular Camilo Cienfuegos is rocky and saline and crops require irrigation as groundwater is not readily available. Drinking water is used for irrigation and this exacerbates existing domestic water shortages experienced by inhabitants, but we could find no evidence of friction between families who grew food and used extra water for irrigation and those who did not. Farmers and their supporters have investigated a number of water recycling and storage schemes, one of the most ambitious being the use of water from the annual draining of the adjacent Panamerican Complex's Olympic Pools. All of these proposals require significant investment in infrastructure and have as yet not been implemented.

A composting scheme has been set up which uses domestic organic waste from individual families for use in urban agriculture. As elsewhere these schemes are voluntary and in 2002 had limited success. A series of workshops was planned to encourage participation.

The Fundacion Antonio Núñez Jiménez has undertaken a detailed study of urban agriculture in the Consejo Popular Camilo Cienfuegos and this has revealed some concerns farmers have with the contractual arrangements they are tied into. For example, farmers have expressed concerns with the quality of seeds supplied to them by the Greater Havana Vegetable Company, a matter over which they have no control given the employment contracts they have entered into. On the other hand these contracts provide farmers with a pension and the Greater Havana Vegetable Company markets some of their produce as well as supplying farmers with additional varieties of vegetable for sale from the farm gate.

Farmers managing the El Pedregal Intensive Cultivation Garden have different attitudes to marketing than those working non-commercial parcelas. The latter have no interest in marketing their produce, while those farming the El Pedregal Intensive Cultivation Garden are active in promoting their goods.

Despite these shortcomings, which can not be considered exceptional, the majority of farmers who had been farming for ten years had the intention of continuing to do so for as long as possible. Furthermore the study found that notwithstanding the less than ideal rocky and saline soil, the yield from Consejo Popular Camilo Cienfuegos sites was 91 per cent of the average yield for equivalent types of urban agriculture in Havana.

ORGANIC URBAN AGRICULTURE

We were interested to find out if organic production was seen as an integral feature of urban agriculture.

Our research revealed that many biological control methods were also proving to be effectively employed on many of the agriculture sites. For example banana stems are grown on many of the urban agricultural plots. Once cut and baited with honey they are extremely effective in attracting insects that otherwise might be tempted by local crops. Such a technique has proved particularly useful in the control of sweet potato weevil.

Another area of bio-development has been the emergence of vermi-compost centres (using worms to promote composting). Whilst many urban agriculture plots undertake their own composting, recently there has been the emergence of sites specifically dedicated to the composting of organic material. Between 1995 and 2002 almost 200 of these units were established with the annual production of organic compost rising from under 3000 to 100 000 tonnes during this period. Other organic farming techniques have also emerged, such as extensive crop rotations, green manuring, inter-cropping and soil conservation.

A conference on urban agriculture hosted by the University of Cienfuegos provided answers to some questions. At the time of our visit, 2002, Cuba did not subscribe to any organic certification scheme. However, by most definitions Cuba's urban agriculture can be considered organic. Since 1989, due to the economic situation, artificial pesticides and fertilisers have been difficult to come by and when chemical treatments are available, they are used in a targeted manner to control specific outbreaks of disease or pests.

The use of organic or natural pest control, for urban agriculture, is explicitly promoted. This has resulted in the accumulation of an extensive body of knowledge, and urban agriculture support workers widely and actively disseminate this to growers.

The development of ecological pest control systems is ongoing. In 2004 new methods for catching small flying insects had been introduced to Cienfuegos. Here the ends of 50 gallon drums and similar metal or plastic disks had been smeared with grease and fixed onto poles. These devices were then 'planted' amongst beds to catch insects, which would otherwise attack crops. This constructed vegetation introduced a layer of surreal poetics to the otherwise utilitarian landscape of urban agriculture – sites began to read like three-dimensional installations by Miro.

Nevertheless, it is important to recognise that, despite great progress, there remain many barriers and difficulties that still need to be overcome. Our site visits revealed farmers who remained sceptical of organic methods. Likewise there are many in the scientific, policy making and government circles that are concerned that new biotechnologies are not developing fast enough. These officials often stated their concern about pests that may defy organic controls.

Much material in this chapter is the result of interviews and conversations with the Cuban urban agriculture specialists, met during field trips to Cuba in 2002 and 2004. We would like to thank the following people for their time and contribution:

Jen Pukonen and Roberto Sanches Medina (Fundación Antonio Núñez Jiménez la Natureza y el Hombre, Havana)

Professor Jorge Peña Diaz (Centre for Urban Studies, CUJAE, Havana)

Professor Socorro Castro and Professor Rene Padron Padron (University of Cienfuegos)

Hector Lopez (Jefe Agriculture Urbana, Province Cienfuegos)

Liliana Mederos Rodrigues (Provincial Delegation for Agriculture: Rodas)

Alena Fernandez and Arura Basques (Urban Agriculture advisors, Rodas and Cienfuegos)

Huber Alfonso Garcia (Delegardo Municipal de la Agricultura)

Pedro Leon Artiz (Jefe Granga Urbana Rodas)

F Gande Iglesias (Jefe Produccion Granga Urbana),

REFERENCES

Bourque, M. and Canizares, K. (2000). Urban Agriculture in Havana, Cuba. *Urban Agriculture Magazine*, Vol. 1, No. 1, 27–29.

Caridad Cruz, M. and Sanchez Medina, R. (2003). *Agriculture in the City: A Key to Sustainability in Havana, Cuba*. International Development Research Centre.

Pretty, J. (2002). *Agriculture: Reconnecting People, Land and Nature*. London: Earthscan.

Ponce, F. (1986). *Sustainable Agriculture and Resistance: Transforming Food Production in Cuba*. Oakland: Food First.

Socorro Castro, A. R. (2001). Cienfuegos, the Capital of Urban Agriculture in Cuba. *Urban Agriculture Notes*, published by City Farmer, Canada's office of urban agriculture. www.cityfarmer.org/cubacastro.html accessed May 2001.

18

URBAN AND PERI-URBAN AGRICULTURE IN EAST AND SOUTHERN AFRICA: ECONOMIC, PLANNING AND SOCIAL DIMENSIONS

Dr Beacon Mbiba

This chapter seeks to identify key aspects of the research record on urban and peri-urban agriculture in Southern and East Africa using a bibliography compiled by Obudho and Foeken (1999) as the starting point and main data base. The dominant status of urban and peri-urban agriculture (UPA) in East and Southern Africa reflects a restructuring (and collapse) of urban economies in the region since the 1970s. This has attracted increased scholarly attention as well as interest from local NGOs and international development organisations that desire to use urban and peri-urban agriculture as an entry point for urban environment management poverty alleviation and food security. However, the success of these interventions will depend on how well we understand both intra-urban agriculture dynamics as well as the context within which it takes place. The latter includes questions of land access and control plus wider urban governance issues. Despite the prevalence of UPA and support from NGOs and donor organisations, local authorities and their planning institutions appear not to consider it a priority.

RESEARCH PATTERNS SINCE THE 1970s: REGIONAL AND SUBJECT FOCUS

Subject focus

Urban and peri-urban agriculture (UPA) research in East and Southern Africa (the region) exhibits several patterns, some of which have been noted before (Rogerson, 2001; Mbiba, 2001). Key among these is the association of UPA with urban economic collapse and poverty. Taking a simple tally of publications by subject and country, based on the bibliographic data of Obudho and Foeken (1999), we note that the earliest studies were in the 1970s. These were in Zambia (1972, 1978, 1979) and Kenya (1977). But they were very few (see Figures 18.1 and 18.2) and focused on UPA as part of the informal sector debate generated by ILO activities at the time.

These were followed in the early 1980s by further studies that had an environmental focus, looking at land degradation, destruction of flora and deforestation. Here, studies in Kenya (Mazingira Institute, 1987) and Zimbabwe (Mazambani, 1982) were notable. These showed that environmental degradation in cities was not only a result of cultivation activities but was also related to construction activities and to fuel and energy needs of the poor populations. Up until the mid 1980s, UPA was not a popular subject at all. In Zambia, where the bulk of the studies was done, the driving force was foreign researchers who looked at theoretical issues, such as the link between UPA and malaria, economic and gender issues.

However, in the early 1990s research output on UPA rose sharply. Economic dimensions and links with poverty and food needs at the household level dominated the work. At the same time, the Swedish sponsored GRUPHEL programme and the Mazingira Institute promoted gender analysis within the UA sector. IDRC promoted a region wide 'Cities Feeding People' research and awareness programme. By 1995, UA was no longer a subject of ridicule in most universities in the region as it embraced other aspects such as food security and nutrition, policy, governance and institutional responses. At the turn of the century, food security and poverty are the key subjects of interest.

Country focus and regional patterns

In terms of countries, although Zambia had an early start, it was soon overtaken by Kenya (1980s) and subsequently Tanzania (early 1990s) and Zimbabwe

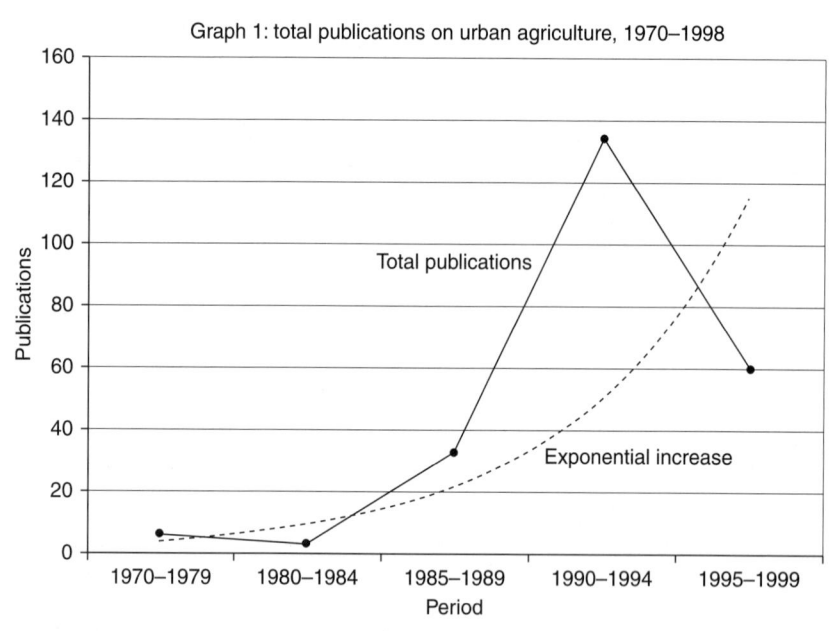

Graph 1: total publications on urban agriculture, 1970–1998

[Graph showing Publications on y-axis (0–160) vs Period on x-axis (1970–1979, 1980–1984, 1985–1989, 1990–1994, 1995–1999), with "Total publications" line peaking around 1990–1994 at ~134 and dropping to ~58 in 1995–1999, and a dashed "Exponential increase" curve.]

Figure 18.1

(mid to late 1990s). As noted above, Zambia's research was dominated by foreign researchers compared to Kenya, Tanzania, Zimbabwe and South Africa where a crop of local researchers competed for research space. However, the more northerly one goes, the higher this research has been donor dependent in terms of research funds. Pure academic research not linked to donor programmes is hard to find except probably in South Africa (see Stren, 1994; Mbiba and Huchzermeyer, 2002) . . .

By 2000, certain features were clear:

- Countries in the South dominate, especially Zimbabwe and South Africa. South Africa is now attracting funds from key donors that are scuttling out of Zimbabwe. Attracting donor funds to Harare, even for a small academic workshop, is now close to impossible.
- Other subjects areas, land, technology, skills training and extension needs are coming to the fore.
- South Africa's output is also on the increase and is likely to be more pronounced as a result of increasing poverty, current positive donor support and a diverse research base in universities in the big cities such as Durban, Cape Town, Pretoria, Johannesburg, Port Elizabeth. Research on technologies, new theoretical interpretations and entry points are likely to emerge here.

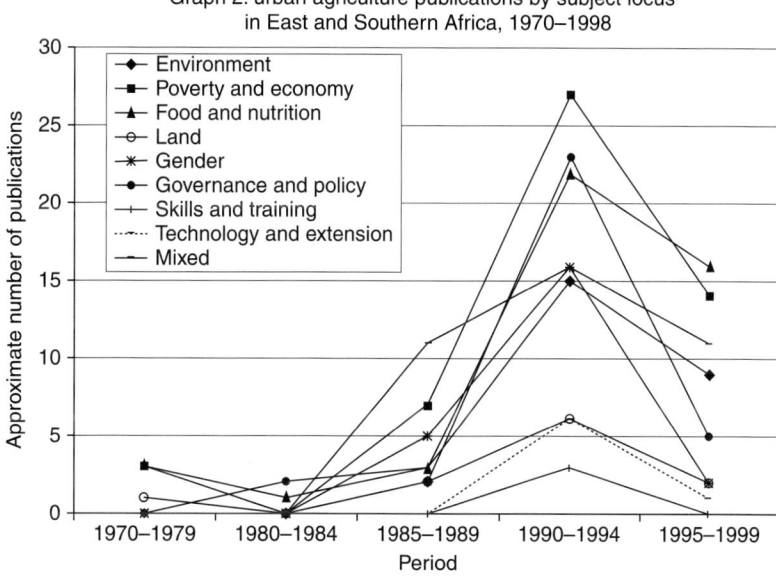

Graph 2: urban agriculture publications by subject focus in East and Southern Africa, 1970–1998

Figure 18.2

- In the greater part of the region, research will remain donor driven in the long term.
- In the smaller countries research output has remained low over the years (Lesotho, Malawi, Rwanda, Botswana, Namibia). In the DRC, Mozambique and Angola, language differences and conflict situations have hindered the sharing of some materials from ongoing research in cities of those countries (see Figure 18.3).

If the Obudho and Foeken (1999) bibliography were to be updated today, certainly the number of publications would be found to be many times more the maximum recorded then. Much of this remains as grey literature in the countries or offices of the donor organisations running programmes in those places.

SUMMARY OF THE CONTEXT OF URBAN AND PERI-URBAN AGRICULTURE IN AFRICA IN CRITICAL RESEARCH AREAS

Although the scholarly work shows that urban agriculture or the growing of food in urban and

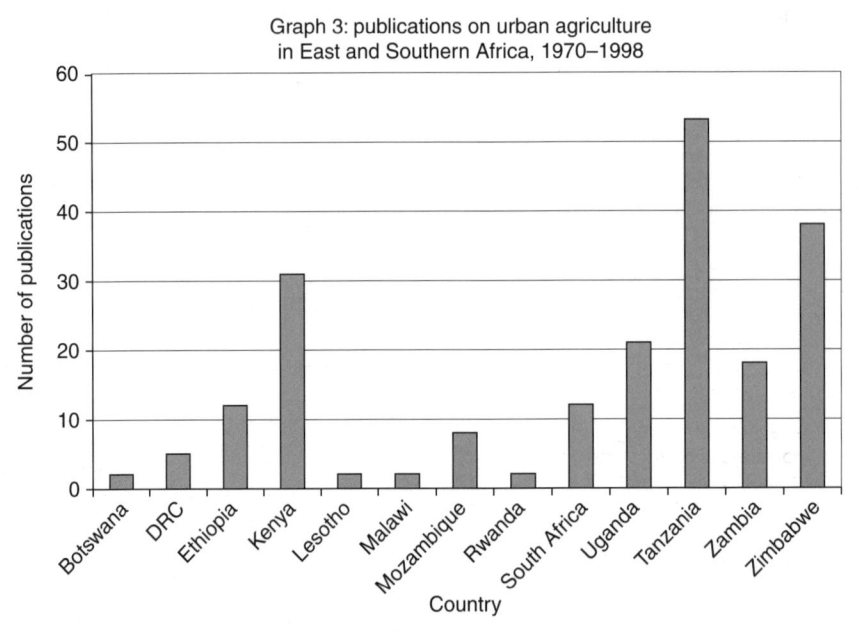

Graph 3: publications on urban agriculture in East and Southern Africa, 1970–1998

Figure 18.3

peri-urban areas (UPA) is not new in Africa (Freeman, 1991; Mbiba, 1995; Grossman et al., 1999; Rogerson, 2001), the phenomenon is now different and requires our attention in several respects. Firstly, since the collapse of formal urban economies in Africa and the debilitating impacts of economic structural adjustment programmes since the late 1970s, both the rich and the poor in Africa's cities have had to seek alternative ways of survival. The new development is that the spatial scale of UPA has grown tremendously and its contribution to household economies has risen to far greater proportions than previously imagined (MDP, 2001).

Probably, the most critical new development arising from the above scenario is the elevation of 'urban agriculture' as a concept in African urban development policy and planning. Never before have there been calls for its inclusion into formal policy. This is associated with growing support from leading international development organisations (such as UNDP, FAO, GTZ, MDP, IDRC and DfID) for the promotion of UA as an element of urban poverty alleviation and sustainable development programmes (see, for example, Bakker et al., 2000; Brook and Davila, 2000; www.ruaf.org; www.fao.org; www.cityfarmer.org).

However, promotion and integration is undermined by a range of contextual, conceptual and institutional conflicts or capacity deficiencies that need to be better understood and responded to. First, is the observation that this new concept of UPA and the activity itself are in practice confronted by an unenthusiastic, restrictive and at times hostile local policy environment. Local settlement policies and institutions remain largely ambivalent to this phenomenon with policy makers unconvinced as to the potential benefits that may arise from its integration into urban development. To an extent, UPA proponents whose global project approach does not capture regional and local diversities, as well as the priority issues and complex demands faced by local policy makers, reinforce the scepticism of policy makers.

Second is the fact that conflicting and contradictory policy/institutional responses are the norm, not only among local authority departments, but also between politicians and technical professionals and among central government ministries. In particular, friction among ministries of local government, agriculture and environment are the norm and reinforce the policy dilemmas at local authority levels.

Third are conflicts related to UPA versus other land uses and users in urban and peri-urban areas especially with respect to land resources. The proponents of urban and peri-urban agriculture have failed to recognise that the success of this activity and its formal acceptance depends on the context and diversity of local dynamics, especially contests for control and access to land resources.

Presently, this access is characterised by conflicts that manifest in *inter alia*, social tensions, destruction of property and the environment, administrative disputes, physical confrontations and loss of economic productivity. Unfortunately, existing policies (mainly town and regional planning instruments) and institutions are ill-equipped to resolve these conflicts in a manner that is sustainable and of benefit to local communities. In some cases, policies and institutions for urban and peri-urban development are non-existent.

Fourth is the question of economic opportunities arising from UPA. Contrary to research evidence, integration of UPA is dominantly conceived of as a domain of the poor. This underplays the untapped and emerging role of entrepreneurial, or for the market, UPA in which the 'elite' and private sector participation has potential to create more jobs as well as broaden the revenue base for urban local authorities.

For most cities in Africa, UPA in the form of horticultural production is not part of the traditional farming systems. Yet if adopted on a wide scale, it would improve unit productivity of land and probably diffuse some of the land conflicts that prevail. Lack of appropriate policies leads to loss of economic potential on one hand and failure to resolve conflicts between small holders versus large international actors, as well as between land demands by this sector versus those of other urban expansion needs. Under a context of globalisation, these issues need clearer policy attention.

ACTORS AND INSTITUTIONS

The rise of urban agriculture as a component of development policy has been noted above. This is associated with poverty alleviation projects and has drawn in all key international development agents although few are at present clear as to how urban agriculture can be turned into a viable development programme. In Tanzania, GTZ has supported the Dar es Salaam Urban Vegetable Promotion Project for many years and has collaborated with other donors to host conferences and publicise urban agriculture (see Bakker et al., 2000). In its attempt to find a role for urban agriculture, the FAO has recently supported a number of expert group seminars in Southern Africa. In collaboration with the Municipal Development Programme, FAO Southern Africa has invested resources into a study of the linkages between land tenure and urban agriculture in Malawi, Kenya, Tanzania and Uganda. Probably the most significant actor is the IDRC that has supported research since the 1970s. This research has been done in collaboration with local partners that included university scientists, local NGOs such as ENDA Zimbabwe, and more recently a combination of different institutions. With the Harare based Municipal Development Programme for East and Southern Africa, IDRC has launched a new initiative on the political economy of urban agriculture in the region. It also supports and works in collaboration with the Consultative Group on International Agriculture Research (CGIAR) and the Nairobi based International Potato Centre (CIP) to co-ordinate research on urban agriculture in Africa.

In every country there is a range of local groups working with urban agriculture practitioners. With the increased attention from international donors, local institutions compete for expected donor resources and time is invested in preparation of project plans to suit frameworks of the donors. On the research front, although there is increased activity, the bulk of the output is for the consumption of the donors and not for publication in traditional international journals. The incentives for the latter on the part of local researchers are diminishing.

CONCLUSION

This chapter reviewed the research record on urban and peri-urban agriculture in East and Southern Africa whose prevalence has increased with increasing levels of economic hardship in every country. In the early years, it was driven by environmental concerns but more recently, economic and food security concerns predominate. Urban agriculture is mainly a subsistence activity as ordinary residents use urban space to produce food under harsh economic conditions. This has raised complex and unresolved issues of legitimacy, land access, tenure and planning.

Therefore, there is a need to understand and elevate development debate on the concept of urban and peri-urban agriculture; on the nature of peri-urban land conflicts and within this to identify innovative policy intervention that benefits local communities and the poor in terms of food security, nutrition and jobs. Clearly, the urban agriculture focus in Africa at the moment does not have an urban ecology, design and regeneration focus like that found in Europe.

There is also a need to understand the nature of current urban development policies with a view to locate where and where not UPA could be integrated into urban development (both spatial and economic). This should include the structures that may be required to facilitate integration as well as capacity building in local authorities, training and research centres, NGOs and communities. Some of the integration challenges were explored in the *Urban Agriculture Magazine* (notably Volume 4, July 2001).

As noted since the early 1980s, the subject of poverty in Africa has 'pulverised research capacity' in general and that for urban policy and management is no exception. Thus, continuous attention needs to be paid to research and training needs so as to support and lead interventions for sustainable urban development. Our research network PeriNET seeks to address some of these problems (see http://www.pidces.sbu.ac.uk/BE/UES/perinet/).

REFERENCES

Bakker, N., Dubbeling, M., Gundel, S. and Sabel-Koschella, U., de Zeeuw, H. (eds.) (2000). *Growing Cities, Growing Food: Urban Agriculture on the Policy Agenda*. DESE/ZEL, Sida, CTA, GTZ, ACPA, BMZ, ETC.

Brook, R. and Davilla, J. (2000). *The Peri-Urban Interface: A tale of two cities*. School of Agricultural and Forest Sciences, University of Wales and Development Planning Unit, University College London.

Freeman, D. B. (1991). *A City of Farmers: Informal Urban Agriculture in the Open Spaces of Nairobi, Kenya*. McGill University Press: Montreal and Kingston.

Grossman, D., Van den Berg, L. and Ajaegpu, H. (eds.) (1999). *Urban and Peri-Urban Agriculture in Africa*. Ashgate: Aldershot.

Mbiba, B. (1994) Institutional Responses to Uncontrolled Urban Cultivation in Harare: Prohibitive or Accommodative? *Environment and Urbanisation*, **6**, 188–202.

Mbiba, B. (1995). *Urban Agriculture in Zimbabwe: Implications for Urban Management and Poverty*. Avebury: Aldershot.

Mbiba, B. (2001). The Political Economy of Urban and Peri-Urban Agriculture in East and Southern Africa: Overview and Research Settings. *Paper presented at a Regional Workshop Organised by the Municipal Development Programme (East and Southern Africa)*, Bronte Hotel, Harare, 27th February–3rd March 2001.

Mbiba, B. and Huchuzermeyer, M. (2002). Contentious development: peri-urban studies in sub-Saharan Africa. *Progress in Development Studies*, **2** (2), 113–131.

Municipal Development Programme (2001). The Political Economy of Urban and Peri-Urban Agriculture in Eastern and Southern Africa. *Proceedings of the MDP/IDRC workshop*, Bronte Hotel, Harare, 28th February–2 March 2001.

Obudho, R. A. and Foeken, W. J. (1999). Urban Agriculture in Africa: A Bibliographical Survey. *Research Report 58/1999: Leiden African Studies Centre and Nairobi Centre for Urban Studies*.

Rogerson, C. M. (2001). Urban Agriculture: Defining the Southern African Policy Debate. *Paper Presented at an FAO Sub-regional Conference on 'Simple Technologies for Crop Diversification by Small Scale Farmers in Urban and Peri-Urban Areas of Southern Africa'*, Stellenbosch University, South Africa, January 15–18.

Smith, D. W. and Tevera, D. S. (1997). Socio-Economic Context for the Householder of Urban Agriculture in Harare, Zimbabwe. *Geographical Journal of Zimbabwe*, **28**, 25–28.

Urban Agriculture Magazine at www.ruaf.org

ABBREVIATIONS

DfID = Department for International Development, United Kingdom

FAO = Food and Agricultural Organisation, United Nations

GTZ = German Development Agency

IDRC = International Development Research Centre, Ottawa, Canada

ILO = International Labour Organisation

MDP = Municipal Development Programme

RUAF = Resource Centre for Urban Agriculture and Forestry, ETC The Netherlands

UNDP = United Nations Development Programme

19

MOULSECOOMB: DISCOVERING A MICRO-PUL

André Viljoen and Katrin Bohn

MOULSECOOMB: DISCOVERING A MICRO-PUL

Figure 19.1 (photo of Moulsecoomb) shows an allotment site in Moulsecoomb, a suburb on the edge of Brighton, on the south coast of England. The site, on a south facing slope is small, having nine standard allotment plots, with an area of 250 m² each, and an overall site area of about 3200 m².

It is occupied by the Moulsecoomb Forest Garden and Wildlife Project, which serves a number of different groups, including community garden volunteers, unemployed people and young children. The project is self-regulated and has financial support from a number of local organisations (Carter, 2001). This diverse group of users, with different backgrounds, desires and interests, has created a usage pattern by which a network of public spaces has been overlaid onto regular urban agriculture plots. This characteristic of 'overlay' is one of the principal features of Continuous Productive Urban Landscapes and the small size of this site shows overlay at its most intimate scale.

Figure 19.2 shows the original arrangement of allotments at Moulsecoomb, nine equal rectangular plots with narrow footpaths inbetween. Figures 19.3 and 19.4 demonstrate how occupation by people has changed the original arrangement of plots, showing how paths have been extended to encompass parts of the planting beds and provide areas for people to use in other ways. Contemplative, recreational, and

Figure 19.1

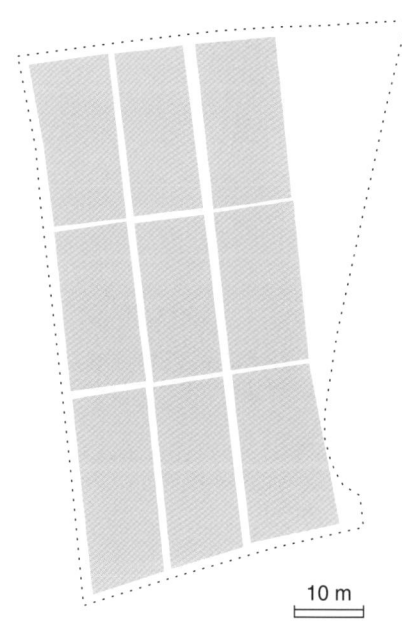

Figure 19.2

Figure 19.1 *Moulsecoomb allotments, Brighton 2002.*
Figure 19.2 *Moulsecoomb allotments Brighton 2002: plots as rented out to occupants.*

private spaces can be found within these extensions to the original footpaths. The whole arrangement can be read as a network (see Figure 19.3) superimposed onto the original layout (see Figure 19.2) resulting in the final mix of occupations, with spaces for growing and living (see Figure 19.4).

Within the overlay of public space, occupants have constructed an enclosure for shelter, and lawns sloping towards the south and the sun, as places for sitting or lying, and a forest garden that offers a retreat. The space fosters social interaction within the local community and between cultivators and other occupants. It can be characterised as a gregarious space, one where social and productive landscapes coincide.

Moulsecoomb represents a micro model for a larger urban intention. It provides a number of clues for how, at the urban scale, Continuous Productive Urban Landscapes may be structured. It demonstrates how a matrix of circulation and occupational spaces can articulate and contain fields for growing. At an urban scale, paths in Moulsecoomb can be imagined as primary pedestrian and cycle ways, planting beds can be thought of as market garden sites, fields in the city. By examing the Moulsecoomb site as found and simultaneously imagining the

Figure 19.3

Figure 19.4

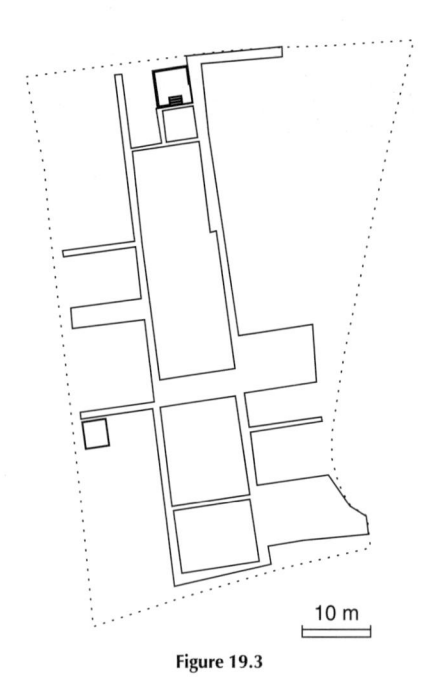

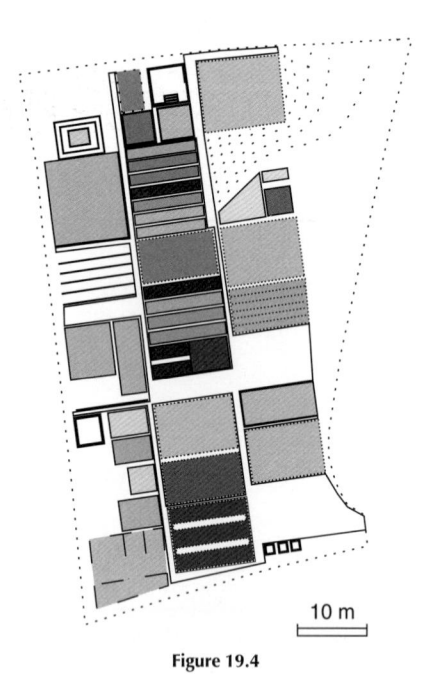

Figure 19.3 *Moulsecoomb allotments Brighton 2002: network of paths and shared space introduced by the occupants.*
Figure 19.4 *Moulsecoomb allotments Brighton 2002: occupation, showing the overlay of paths, shared space and plots for food growing.*

allotments as city blocks or large fields of urban agriculture, a pattern of urban development can be imagined, in which the principle of overlay and Continuous Productive Urban Landscapes has been applied at a number of different scales.

In addition to seeing how occupation modifies a strict municipal geometry by providing space for social life, the site shows productive urban landscapes as a visual resource and marker of seasonal change.

From an adjacent commuter bridge at a railway station, the Moulsecoomb site offers a view to pedestrians. This view is important to our understanding of another aspect of urban agriculture. It demonstrates how agricultural fields provide urban dwellers with a connection to horticulture, to agriculture and by extension to an understanding of human interaction with nature. This connection, which is visual, answers a powerful human desire for contact with the countryside, or at least a need for contact with an idea of nature and countryside. The significance of this desire is demonstrated by the popularity of suburban dwellings on the rural fringe and the popularity, for those who can afford it, of a weekend cottage. For a city dweller, having regular contact with urban agriculture is like having a house in the countryside.

Imagine the qualities brought to a city if the conditions found in Moulsecoomb's productive urban landscape were repeated across a city, to become a Continuous Productive Urban Landscape.

PLANNING FOR CPULs: INTERNATIONAL EXPERIENCE

Characteristics of allotments, distinct edges and teritories, fine grain, direct use of materials, repetition and variation
Figure 19.5

Figure 19.5 *Allotments in Rye, England.*

A network of paths, places for rest and work · photo: Antonia Faust

REFERENCE

Carter, W. (2001). *Seedy Business – tales from an allotment shed.* Moulsecoomb Forest Garden and Wildlife Project.

20

ALLOTMENTS, PLOTS AND CROPS IN BRITAIN

H.F. Cook, H.C. Lee and A. Perez-Vazquez

Figure 20.1 *Urban allotments in London, looking toward the Millennium Dome.*

In Britain, the allotment garden has long been a part of working-class culture, and the practice has survived the widespread availability of cheap food, post Second World War. Today, urban and peri-urban agriculture (UPA) in Britain continues to be useful as a means of supplying some food and financial income to urban people (Garnett, 1996a, b; Rees and Wackernagel, 1996; Dunnett and Qasim, 2000), but it also has many other important benefits:

1. social (leisure, empowering local groups of people such as women, therapy for people with special needs, rehabilitation for young offenders).
2. environmental (renovation of derelict urban sites, diversifying urban land use, increasing biodiversity, reducing the ecological footprint).
3. human (encouraging personal qualities such as altruism, improving quality of life by social contact, health benefits of exercise, better quality and more diverse food intake).
4. economic – stimulating local economies.

This chapter will focus upon UPA in Britain and will consider allotments and smaller, less-regulated urban plots of land. Most information in the literature refers to allotments and this will be the main focus. The key issues that will be addressed are:

1. the organisation of allotments and implications for urban planners.
2. practical production issues, including implications for soil contamination and water pollution.
3. the economic implications of allotment gardening in Britain.

THE ORGANISATION OF ALLOTMENTS

In Britain, allotments are defined as pieces of land in cities and towns provided either by local authorities

Figure 20.1

or landlords where plot holders grow their own food, mainly vegetables and soft-fruits for self-consumption and for their families (House of Commons, 1998). The concept of 'plots' is more recent and usually describes individual parcels of land close to houses in urban areas. They are smaller than allotments, variable in size and utilised by people opportunistically, on an ad hoc basis (Perez-Vazquez, 2000). UPA has special relevance to the present day, as a means of reducing food's 'ecological footprint.' Therefore, allotments are increasingly seen in Britain as an integral and diverse component of the urban landscape in London and elsewhere.

Figure 20.2 *Urban allotments in London, looking towards Canary Wharf.*

Figure 20.2

There are approximately 296 900 allotments in England across 7800 sites, covering a total of 10 290 hectares. This equates to one allotment for every 65 households (NSALG, 2000). Prime examples near London include the Lee Valley and the 'green finger' area along the Hogsmill river south-west of London (NRA, 1994). Urban food growing in London covers about 30 000 active allotment holders gardening a total of 831 hectares. In London there are also about 77 community gardens and 18 city farms, school gardening projects and countless individuals who grow food in small plots – from a few herbs and a tomato plant upwards – in their back gardens, see: http://www2.essex.ac.uk/ces/ConfsVisitsEvsGrps/LocalFoodSystems/localfoodtg.htm

Allotment size

Allotment sites vary in terms of their size, services and facilities provided. The average size of the allotment is ten rods or 300 square yards (250 m^2) (Crouch and Ward, 1988; Perez-Vazquez, 2000). This plot size was originally set by law, to meet the needs of households and not for trade or other businesses (Blackburn, 1998). However, demands for increased housing and commercial capacity in urban areas has led to land scarcity for allotment provision (Perez-Vazquez, 2000). This has resulted in reduced allotment size (Radice, 1997) and some people in urban areas have had to try and utilise much smaller parcels of land (plots) in a less-controlled way than for allotments. Thus, space-demanding crops such as potato are not cultivated in many urban plots (Perez-Vazquez, 2000). This is especially true for early potato production, planted in February and harvested in July, which would conflict with other early season vegetable and salad crops. The traditional concept of self-sufficiency is therefore much less relevant for modern urban plots. Indeed, demand for allotments has increased during the 1990s and more people have been accommodated by reductions in average plot size. For example, in the Borough of Haringey, London between 1971–1987 the number of allotment users increased by one-third and were accommodated by reduced plot sizes (Radice, 1997). At Imperial College at Wye, a participatory research project was carried out in three different locations in the southeast of England (Perez-Vazquez, 2000). At each site, varying numbers of allotment holders were interviewed (74 at greater London, 34 at Ashford and 19 at Wye, both in Kent) in order to analyse allotment

Figure 20.3 *Floral allotment.*

management and use. Participatory methods were used, including:

1. semi-structured interviews with allotment holders and key informants
2. mapping allotments
3. time lines – a description of major events in the period of ownership of the allotment
4. seasonal calendars with main activities highlighted.

Results indicated that most urban allotment users rented their land and therefore had almost no control over plot size allocated to them, though individuals sometimes rented more than one plot or even a fraction of a plot. The intended purpose of the allotment (growing food *versus* hobby or leisure) was also found to influence size: whether more than one allotment, or a fraction of an allotment was utilised. The personal characteristics of the user were also found to affect chosen allotment size, such as time available, level of fitness and strength of commitment. It was concluded that allotment size and their provision should consider the local demand (waiting list) and also that about 11–21 per cent of people who live close to allotment sites are normally interested in participating (Perez-Vazquez, 2000).

Allotment plot design

The design of an allotment plot needs to take account of its intended purpose: therapeutical, hobby or recreational, commercial, self-consumption, or mixed purpose. Allotment gardening is one of the most popular leisure or hobby pursuits in Britain (Crouch and Ward, 1988; Garnett, 1996a). Indeed, allotment management is considered by many people as a leisure activity rather than as a means of growing food (Thorpe, 1975). Thorpe has mentioned that the word 'allotment' should be

Figure 20.3

replaced by the concept of 'leisure gardens' because the former has an historical stigma of low income and relative poverty. Additionally, allotment site design should consider not only the individual plots but also communal compost areas, shed(s), recreational areas, and even some occasional space for orchard and woodland. In addition, allotment sites should be strategically located close to demand and as far away from known sources of contamination as possible, such as old railways, bomb sites and some industrial brownfield sites (Perez-Vazquez, 2000).

Gender

Information from the participatory studies and semi structured interviews described above (Perez-Vazquez, 2000) suggest that women grow more flowers as decorative plants on allotments and are not likely to follow classic rules of planting. Herbs, flowers, vegetables and potatoes are likely to be equally important for them. Many women tend to prefer an informally managed plot. Some female allotment holders are reluctant to kill even weeds and pests such as slugs. By contrast, most male

Figure 20.4 *Crop rotation design as managed by a female allotment holder (Perez-Vazquez, 2000).*

holders tend to give priority to potato, onion and soft-fruit production. They tend to grow plants in straight lines and keep plots neat, and free of weeds and pests (Perez-Vazquez, 2000).

PRACTICAL PRODUCTION ISSUES

The choice of crop species, varieties and their likely performance

For both genders, the growth of vegetables on allotments is controlled by the desire to grow potatoes. There is a key requirement to avoid repeated growth of potatoes on the same ground each year, to avoid the build up of two major pest species of potato cyst nematode (*Globodera rostochiensis* and *G. pallida*) (Winfield, 1990). This necessitates the development of a crop rotation on most allotments (Perez-Vazquez and Anderson, 2000), such as the example in Figure 20.4.

There is generally an increasing diversity of crop species and varieties within species seen on allotments in Britain, reflecting not only gender differences but also an increase in the ethnodiversity of participants (Garnett, 1996a; Perez-Vazquez and Anderson, 2000). The use of greenhouses or simple cloches means that people can grow food in allotments almost all year round, or at least start earlier each season, usually with salad crops (Perez-Vazquez, 2000). Generally there is a lack of practical information on crop species and variety choice for allotments and how they are liable to perform in urban conditions, i.e. higher temperatures, more shading, and variable soil factors such as fertility, compaction and the presence of some soil contaminants. An example of the range of crop species and varieties chosen is shown for one site in Table 20.1.

The yields of crops grown on allotments tend to vary and even more so for smaller plots of land close to houses (Perez-Vazquez, 2000). This variability is thought to be due to the lack of high yield as a priority, with many people growing such crops for recreational reasons (Gilber, 1989; Perez-Vazquez, 2000). Often, an allotment with a diverse array of small areas of vegetable and fruit species will tend to give lower yields per species than one where a smaller range of species are grown but on a larger scale. For example, potatoes are sometimes grown as a major crop in well-organised rows (giving higher yields) or occasionally as a recreational crop in just one or two short rows (giving lower yields). Allotment holders also tend to choose a relatively wide range of crop varieties (Gilber, 1989; Radice, 1997). Such diversity can be due to social factors such as family size, food preferences, and level of commitment (Perez-Vazquez, 2000). Thus, one urban allotment plot may show relatively large yield variation compared with a rural equivalent. One example for an urban allotment plot is shown in Table 20.1.

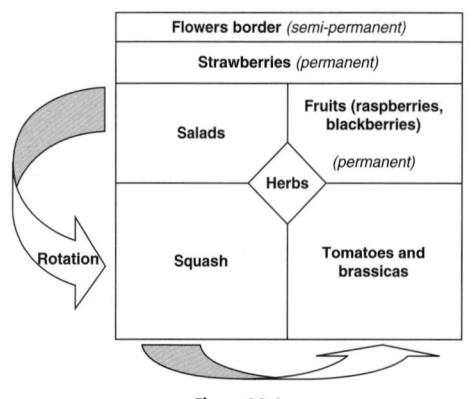

Figure 20.4

Figure 20.5 *Harvesting potatoes on an urban allotment.*

Figure 20.6 *An example of crops growth in sunken pots to make the best use of applied water.*

Water issues

The irrigation of UPA in Britain depends mainly upon harvested rainwater (from the roofs of sheds into containers) and also mains supplies. The application of irrigation water can range from the use of hosepipes, to sprinklers, to hydroponic production systems. At almost every allotment site in Britain there is a water supply and water tanks are eventually provided. Whilst there is normally free access to this water, only watering cans may be used: the use of hose pipes is restricted or even prohibited in most allotments sites, though this does vary, see http://www.sags.org.uk/MerlinTrustReport.php4/index.php, or http://www.inthelimelight.co.uk/localgov/allotments/crops_water.html. Carpets, horse manure and mulches are used to retain soil water and suppress weeds. Pots may be buried in the soil to allow water to be retained more effectively in the vicinity of plant roots. Watering is usually difficult in summer time, especially for the elderly and less fit, due to the prohibition on use of hose pipes.

UPA and water pollution

The possibility of UPA production presenting a hazard to natural waters depends upon the kind of management, and also the local 'environmental vulnerability,' of natural waters. For example, significant areas where an economically major aquifer is overlain by soils of high or intermediate permeability occurs within both Greater London (NRA, 1994) and throughout north and east Kent (NRA, 1994). Allotment sites in the Ashford area are over aquifers from which public water supplies are drawn. Whilst the risk from heavy metals is discussed below, major risks to waters (both groundwater and from surface runoff) could arise from excessive use of pesticides and from nitrogen fertilisers, both artificial and organic. The polluting effects of urban cultivated areas adjacent to watercourses and effects on the underlying hydrogeology are potentially important, but largely ignored.

Figure 20.5

Figure 20.6

Table 20.1. *Species, varieties of vegetables and their yields obtained from allotment holders in Ashford and Wye, Kent (Perez-Vazquez, 2000)*

Vegetables (varieties italicised in brackets)	Ashford, Kent (1999) lbs (kgs)	Wye, Kent (1999) lbs (kgs)	Wye, Kent (2000) lbs (kgs)	Average yield 250 m² lbs (kgs)
Potatoes				
Early (*Maris bard; Pentland Javelin*)	65 (29.5)	40 (18.1)	60 (27.2)	57 (25.9)
Second early and main (*Wilja, Maris peer; Pentland crown; Nadine; Desiree*)	180 (81.7)	200 (90.8)	280 (127.1)	220 (99.8)
Beans (*Pots*)				
Broad (*The Sutton*)	—	25 (11.3)	80 (36.3)	52.5 (23.8)
Runner (*Enorma; Desire*)	77 (34.9)	40 (18.1)	10 (4.5)	42 (19.2)
French (*Daisy*)	25 (11.3)	56 (25.4)	16 (7.3)	32 (14.5)
Onions				
White (*Bedford champion; Dutch yellow; Ailsa Craig*)	68.5 (31.1)	20 (9.1)	30 (13.7)	39.5 (17.9)
	14 (6.4)	18 (8.2)	—	16 (7.3)
Spring (*Paris; Barletta*)	9 (4.1)	—	53 (24.0)	31 (14.1)
Brussels sprouts (*Cascade F1; Widgeon*)	35 (15.8)*	20 (9.0)	—	55 (12.4)
Cabbage (*Celtic; Jupiter*)	55 (25.0)	30 plants	10 plants	15 plants (24.9)
Carrots (*St Valery; James Scarlet*)	92 (41.7)	30 (13.6)	5 (2.3)	42 (19.1)
Swede (*Marian; Purple top; Magnificent*)	7.2 (3.3)	14 (6.4)	5.4 (2.5)	8.9 (4.0)
Beetroot (*Detroit 2; Monodet*)	24 (10.9)	8 (3.6)	6.5 (3.0)	12.8 (5.8)
Cauliflower (*Canberra; Snowball*)	50 (22.7)	20 plants	—	35 (15.9)

Table 20.1 (continued)

Vegetables (varieties italicised in brackets)	Ashford, Kent (1999) lbs (kgs)	Wye, Kent (1999) lbs (kgs)	Wye, Kent (2000) lbs (kgs)	Average yield 250 m² lbs (kgs)
Lettuce (*Little gem; Dolly; Salad bowl*)	—	80 plants	45 plants	62 plants
Leek (*Argenta; Catalina*)	27 plants	17 plants	35 plants	26 plants
Peas (*Rarly Onwardand Hurst Green shaft*)	24.5 (11.1)	18 (8.2)	3.5 (1.6)	15.3 (6.9)
Radish (*Scarlet globe; Prinz rutin; Sparkler*)	6 bunches	20 bunches	6 (2.7)	11 bunches
Tomato (*Marmande; Alicante; Golden Sunrise*)	32 (14.5)	24 (10.9)	27 (12.3)	27.6 (12.5)

1 lb = 0.45359 kg; * = estimated yield available only.

Soil contamination

Contaminated land is usually defined as 'land that contains substances which, when present in sufficient quantities or concentrations, are likely to cause harm, directly or indirectly, to people, to the environment, or, on occasion, to other targets' (Garnett, 1996c). If land is sufficiently contaminated, handling soil, or breathing vapours emitted from such land, or eating contaminated food grown on such land can pose significant health risks. The key issue is that much urban land remains unsurveyed and potential practitioners of UPA are therefore unaware of risks, real or potential. In some situations, concerns about land contamination may dissuade people from starting UPA. Some surveys have already suggested that lead (Pb) concentrations in garden-grown crops can be higher than in rural areas in England (Davies et al., 1983), and in Germany (Alt et al., 1982). However, not all results are alarmist: a study in the Netherlands suggested that cadmium (Cd) and Pb levels in soils and crops from allotments were more typical of national levels (van Lune, 1987). Moir (1985) found that seven out of eight allotment sites in Greater London were contaminated with Pb, that the soil metal content decreased with distance from central London and that spinach accumulated significantly higher Pb, zinc (Zn) and Cd levels than lettuce and radish. In Britain, legislation from 1 January 1986 to reduce permitted Pb concentrations in petrol from 0.40 to 0.15 g/l, has clearly led to a reduction in Pb concentrations in air (Denton, 1988). However, there appears to have been limited assessment of levels in urban soils and crops since then. A programme of urban land assessment therefore needs to be organised to quantify both the risk of land contamination and that of drainage water.

THE ECONOMIC IMPLICATIONS OF UPA

It has to be recognised that urban agricultural production is undertaken in a different way from that on farms (Smit et al., 1996), so that standard, accepted indicators of profitability are frequently not appropriate. Whilst British allotment production was traditionally a means of providing food for families on minimal incomes (Crouch and Ward, 1988) it is today more of a leisure pursuit (Perez-Vazquez, 2000). However, this does not prevent allotment holders from being aware of profitability as an issue. For example, in Britain it has been shown that growing brassicas, tomatoes, leeks and onions gives better financial returns for labour than potatoes and legumes, which are usually cheap and plentiful in supermarkets and shops (Riley, 1979). This therefore means that allotment vegetable production is often more concerned with high value crops rather than with dietary value. However, allotment holders are also motivated by a concern for food quality, often preferring to grow their own fruit and vegetables organically and therefore having more confidence in their freshness, flavour and nutritional value (Perez-Vazquez, 2000).

It is therefore proposed that a financial analysis of allotment gardening should allow for the following points:

1. Fixed costs such as labour may not be relevant, since leisure is such an important objective and, in any case, the selling of produce from allotments is not allowed by law. The inclusion of labour costs in a financial analysis of allotments is therefore likely to conclude that they are not profitable.
2. Age of the allotment enterprise is important, since newcomers always incur higher initial costs due to investment in a shed, tools, timber and other items of infrastructure.
3. Proximity of the participant to his/her allotment can be important, affecting costs of transportation.
4. The type of allotment holder is significant (such as retired, disabled, elderly). For example, retired people are not obliged to pay rental in many Council allotment sites for their plot (House of Commons, 1998), yet they are known to be more likely to purchase and use chemicals (Perez-Vazquez, 2000).
5. The growing system, especially whether it is organic or non-organic.
6. If growers value leisure and contribution to quality of life, this may need to be incorporated in an economic assessment.

Actual studies on the profitability of urban plots are scarce. One potentially useful survey is now out of date (Best and Ward, 1956) and so not reported here in detail. A more recent case study, (Perez-Vazquez, 2000) monitored the estimated average annual total market value of fruit and vegetables per urban plot in Ashford, and Wye, in the UK, by considering the current retail value for conventional food from two local supermarkets (Sainsbury's and ASDA). The estimated average annual total market value of the produce per plot was £462. To determine the net average value, various costs needed to be determined and deducted. The internal economic costs included 'fixed' items such as land rental and 'variable' items such as costs of plants, seeds, tools, equipment, and petrol for transport. Determining such costs was complicated by the clear difference between more recent and more established allotment holders: the former always incurred high initial start up costs. Also, all allotment holders were able to demonstrate optional cost savings, such as the use of saved seeds, and walking or cycling to the allotment rather than using a car. It was also considered whether the external economic costs of allotment activities should be estimated. These could include factors such as remediation, to cover the environmental costs of: a) excessive fertiliser or agro-chemical use; b) air pollution due to use of cars for transport. However, such environmental costs are likely to be lower than those associated with imported produce and so were not estimated. For the two allotment sites above studied by Perez-Vazquez (2000), the average net value of produce per plot of 120 m^2 was therefore estimated to be £325, or £677 for a standard 250 m^2 urban allotment. Under this type of financial analysis (not including labour), allotment gardening appears to be profitable, particularly for those people who grow and produce organic food. The inclusion of labour is extremely difficult, since allotment holders invest a wide range of time which depends upon age, motivation and many other social factors discussed above. It seems likely that, if labour were to be built into such financial assessment, then allotment production would be deemed unprofitable. However, many allotment growers view their production as part of a leisure activity and not relevant to financial analysis (Perez-Vazquez, 2000).

CONCLUSIONS

1. Allotments as a component of UPA in England have many roles for people including diet, pleasure, relaxation, healthy exercise, culture, friendship and the encouragement of a sense of community. Urban allotments and plots are therefore much more about lifestyle than about self-sufficiency in food and/or saving money.
2. Technical aspects of the production of fruit and vegetables on allotments and plots are poorly understood, including crop and variety choice and rotation design. Britain's culture is now much more diverse than previously and the implications of this for allotment planning and management are poorly understood. We also know little about the effects of former urban land use, and possible soil contamination, on the safety of UPA and likewise the potential impact of allotments on local water quality.
3. Allotment gardening does not seem profitable, particularly if time/labour are included in the financial analysis. However, it is argued that profitability is not a good indicator of the value of urban plots because economic assessment does not easily incorporate the many important social factors.
4. It seems likely that urban allotments and plots will continue to be important within cities, towns and villages, with the traditional ageing male allotment holder giving way to other groups. Whilst the future role of allotments as agents for food production may be nationally insignificant, their value in contributing to human well-being, especially in urban areas, will be much greater and should be valued as such.

REFERENCES

Alt, D., Sacher, B. and Radicke, K. (1982). Ergebnis einer Erhebungsuntersuchung zur Nährstoffversorgung und Schwermetallbelastung von gemüsebaulich genutzten Parzellen in Kleingärten. *Landwirtschaftliche Forschung*, Sonderheft **38**, 682–692.

Crouch, D. and Ward, C. (1988). *The allotment, its landscape and culture*. Faber and Faber: London.

Davies, B. E., Davies, W. L. and Houghton, N. J. (1983). Lead in urban soils and vegetables in Great Britain. In *Heavy metals in the environment*. Proceedings of International Conference, Sept 1983, Heidelberg, Germany, Vol. II, pp. 1154–1157. CEP-Consultants: Edinburgh.

Denton, D. (1988). Lead in London soils and vegetables. *London Environmental Supplement*, No. **16**, 1–11.

Dunnett, N. and Qasim, M. (2000). Perceived benefits to human well-being of urban gardens. *HorTechnology*, **10** (1), 40–45.

Garnett, T. (1996a). *Growing food in cities: A report to highlight and promote the benefits of urban agriculture in the UK*. National Food Alliance and SAFE Alliance Publications. London, UK.

Garnett, T. (1996b). Farming the city: the potential of urban agriculture. *The Ecologist*, **26** (6), 299–307.

Garnett, T. (1996c). Harvesting the cities. *Town & Country Planning*, **65** (10), 264–266.

Gilber, O. L. (1989). Allotments and Leisure gardens. In *The Ecology of Urban Habitats* pp. 207–217, Chapman and Hall: London–NY.

House of Commons (1998). *Fifth Report: The future for allotments*. Volume I. Report and Proceeding of the Committee. Environment, Transport and Regional Affairs Committee Environment Sub-Committee. Session 1997–98. London. p. 61.

Moir, A. M. (1985). An investigation into contamination of soils and vegetables from gardens and allotments in Greater London. MSc Thesis. Imperial College of Science and Technology. Centre for Environmental Technology. p. 163.

National Rivers Authority (1994). The Wandle, Beverley Brook, Hogsmill Catchment Management Plan Consultation Report. National Rivers Authority, Thames Region: Frimley, UK.

NSALG (2000). *Joint Survey of Allotments in England*. National Society of Allotment and Leisure Gardeners Limited, O'Dell House, Hunters Road, Corby, Northants NN17 1JE.

Perez-Vazquez, A. (2000). The Future Role of Allotments in Food Production as a Component of Urban Agriculture in England. Final Report to Agropolis-IDRC. Imperial College at Wye, Ashford, United Kingdom.

Perez-Vazquez, A. and Anderson, S. (2000). Urban agriculture in England, Perspectives and Potential. In *CD Proceedings of the International Symposium: Urban Agriculture and Horticulture; the linkage with urban planning* (Hoffmann, H. and Mathey, K. eds.). Humboldt University of Berlin and TRIALOG, Berlin, Germany.

Radice, D. E. (1997). *Allotments: food, plant diversity or both?* MSc Thesis. Imperial College of Science Technology and Medicine, Centre for Environmental Technology, London, UK, p. 116.

Rees, W. E. and Wackernagel, M. (1996). Urban ecological footprints: Why cities cannot be sustainable – and why they are a key to sustainability. *Environmental Impact Assess Review*, **16**, 223–248.

Riley, P. (1979). The allotment campaign guide, *Friends of the Earth*, p. 63.

Smit, J., Ratta, A. and Nasr, J. (1996). *Urban agriculture: food, jobs and sustainable cities*. United Nations Development Programme (UNDP). Publication Series for Habitat II. Volume one. p. 302.

van Lune, P. (1987). Cadmium and lead in soils and crops from allotment gardens in the Netherlands. *Netherlands Journal of Agricultural Science*, **35**, 207–210.

Winfield, A. L. (1990). Potato cyst nematodes. In *Crop Protection Handbook – Potatoes* (J. S. Gunn, ed.) pp. 89–94, British Crop Protection Council: Farnham, UK.

21

URBAN FOOD GROWING: NEW LANDSCAPES, NEW THINKING

Simon Michaels

FOOD GROWING IN URBAN AREAS

Urban areas, especially in the UK, are typified by their organic growth which has resulted in a diverse patchwork of public and private open spaces. The design and management of these spaces depends on many factors. Whilst many areas have been designed and continued to be managed in a positive manner, other spaces are 'left behind' in terms of a clear sense of ownership and responsibility. Finding positive uses for these spaces has been one of the challenges for urban planning in the late twentieth century, with an increasing number of projects now including an element of food growing.

The benefits of these urban food growing projects can be multifaceted. They include improvements to the environment and landscape setting of developed areas, as well as significant socio-economic benefits including health and community development.

Despite these proven benefits, the perceptions of the positive impact of urban food growing on the character and quality of the urban landscape, do not appear to be shared by all. The landscapes created as a result of urban food growing projects tend to share some common characteristics, which sit outside of the dominant approaches to the design of the urban environment by landscape and urban design professionals.

These characteristics include a fine grain, an introverted and often unplanned character, and landscapes which are often in a state of change. To some, they are perceived as detractors to the quality of urban landscape. But to others they are some of the most well loved, democratic, and useful landscapes.

Urban food growing projects tend to be an expression of grass-roots activity, led by local people. They often provide an important source of fresh food for families with poor access to affordable fresh produce, or may be developed in order to meet other social needs. The driver for the development of these projects is, therefore, often not one of environmental improvement, nor are such projects necessarily developed with the assistance of planning or design professionals.

How can food growing be accommodated in urban areas in a way which is acceptable to those responsible for planning and design considerations? What needs to change? Is it in the way in which spaces dedicated to food growing are designed, or does it require a change in the mindset of design and planning professionals?

THE LANDSCAPE CHARACTER OF URBAN FOOD GROWING PROJECTS

Spaces which accommodate urban food growing projects take on many different forms. They include the allotments of middle England, the rooftop gardens in Russia, and the Cuban vegetable plots. They may be located on large tracts of open space where, for example, one can find allotments and community orchards, they may be one component of a community garden, or they may include small patches of space in gardens or even window boxes.

The opportunity for food growing in urban areas represents a huge untapped resource. It has been shown that the amount of space available in urban backyards in Vancouver, for example, is equivalent to the amount of active farmland in the province. Whilst it is of course highly unlikely that all such spaces would be used for food growing, there remains a huge potential for food growing at the various scales and using different models of ownership and management.

Despite the diversity of urban food growing projects, there are some common threads which can be identified, in terms of the planning, design and appearance, and impact of food growing on the character of urban landscapes:

- an intensity of activity – food growing is generally a labour intensive activity; urban agriculture creates a peopled and well-loved landscape.
- a changing character – the inter-weaving patterns of cultivation and cropping, linked to the seasons, create landscapes which always change and evolve.
- introversion – the focus is on the business of growing; not outward focused or integrated within the landscape context.
- an unplanned, make-do culture – food plots tend towards the untidy with the use of recycled elements to create temporary or semi-permanent structures.
- greenness – the verdant and vibrant appearance of a healthy plot contrasts strongly with many other forms of municipal landscape.
- a fine grain – the patchwork of small-scale growing and the highly personal character of most urban food projects creates a fine grain or urban texture.

Another common theme relates to location: food growing areas in the public realm are often in leftover spaces, or on land unmanageable by other means. Environmental improvement schemes in public sector housing estates, for example, often need to identify new uses for large open areas, which are performing no useful function. One solution to these types of under-used spaces is to create community gardens, such as those developed through the urban agriculture initiatives in the Metropolitan Borough of Sandwell. These types of project have shown the weaknesses in an approach to urban planning which is based on a designer focused, rather than the people focused, model of urban life. The post-war ideal of space flowing around high-rise buildings has been discredited for several decades, but is only slowly being replaced by an approach to urban planning which offers ownership and control to local people to manage their environments as they choose.

At the other end of the scale, food growing can fit into any unused corner in the ground, containers, or on buildings. Back gardens already produce a significant proportion of seasonal fruit and vegetables for families in the UK. Food crops in back gardens in urban areas also create a significant contribution to the biodiversity of towns and cities.

Approaches to the layout and management of food growing in urban layers also shows great variation, from tidy rows and raised beds, permaculture plots and orchards, and ornamental plots and portages. The inventiveness and opportunity for personalisation in the design and management of food growing areas is one of the most endearing characteristics. No two allotments are quite the same, and no two gardeners can ever agree on the best way forward. Creating urban landscapes which can accommodate this very personal and human interaction, is almost impossible with traditional approaches to urban master-planning. Community gardens, vegetable plots, and other models of urban agriculture, enable this people-centred approach to the management of urban spaces in a remarkable way.

Where does this fit with current thinking on urban landscape?

Responsibility for the planning and management of urban landscape falls under several different organisational remits. These include:

- local authority planning departments – a land-use remit.

- local authority or private sector landscape architects and urban designers – a design remit.
- local authority parks or landscape maintenance departments – a management remit.
- local authority Agenda 21 officers – a community remit.
- activities of special interest groups and community organisations – focused or single issue remits.
- private areas – a management remit with diverse objectives.

These remits overlap, yet rarely co-ordinate smoothly, especially around issues of control and ownership, and in consequence, funding. Still dominant in most urban areas is a top-down, conventional approach to land-use planning.

The successful models of urban food growing in the UK almost all rely on the energy of committed local groups, with help from a public sector starved of cash and time. Until local power sharing becomes embodied into land-use and open space strategies, and the aesthetic of food projects is more accepted, urban food growing will struggle to find anything but a marginal existence.

The key ingredient – people

Food growing projects have a huge power to bring people together and engender a lost sense of community. They act as a resource for learning, an opportunity for minority and special needs groups, and can contribute to local economic development. They help people to connect with and care for their local environment.

These projects help counter vandalism; increase the surveillance of urban areas; create real health benefits; and offer a cost-effective way in which to manage under-used spaces.

Resources are well spent where targeted to enable people to take control of their local environment. This may include educating people in food growing techniques and related activities such as cooking skills; and facilitating the design and implementation of landscape projects.

Local distinctiveness

The landscapes which urban food growing generates are intrinsically embedded in their local context. Plants which are grown are those which will survive in the local microclimate; local varieties may still be used; seasonality is appreciated; plants grown may reflect cultural preferences.

Biodiversity is enhanced through the diversity of habitat and the concentration of growing activity.

Community-led projects are often inventive in their design and materials and work with recycled components from the local area.

The result can offer a new paradigm – a people-centred landscape contrasting with the blandness of municipal landscaping and juxtaposed with 'designed' urban landscape.

CONCLUSION

Urban agriculture initiatives take many forms. Most are typified by their changing appearance and fine-grained, people-centred character. This characteristic challenges planning professionals to review their perceptions of what constitutes good urban design, so that they might encourage the development of food growing projects into the urban fabric. If this can be achieved, richer, more diverse townscapes will emerge.

22

PERMACULTURE AND PRODUCTIVE URBAN LANDSCAPES

Graeme Sherriff

Permaculture is all about solutions for sustainable living. It is an approach and methodology with a strong scientific basis and ethical justification, and it is already strongly represented in local food, Agenda 21 and Green political circles. This chapter looks at permaculture and its relevance to urban agriculture. It hinges on two questions: what is permaculture; and how can permaculture inform urban agriculture? Or, to put it another way, should the urban agriculturist also be a permaculturist?

PERMACULTURE

Permaculture is about producing food in an environmentally-sound way. It's about people growing their own food on their own land and using it for themselves, their immediate family and possibly the local community. This is a very crude description of what Bill Mollison, permaculture's founder, envisaged in the 1970s. The name itself derives from his vision of **perm**anent agri**culture** (Mollinson, 1992).

A definition provided by a contemporary permaculture designer brings us more up to date: 'Permaculture has evolved into a system for the conscious design of sustainable productive systems which integrate housing, people, plants, energy and water with sustainable financial and political structures.' (Hopkins, 2000, p.203)

Since its inception, permaculture has developed and diversified. The key words in Hopkins' definition are 'design', 'sustainable' and 'productive systems'. The key characteristic is that it sets out to maximise the beneficial relationships through the effective placement of elements. Bill Mollison devised a list of key principles by which this is effected, which are based on the example of nature. These are espoused in his 1988 'Designer's Manual' (Pepper, 1996):

1. Work with nature.
2. All nature plays a part in working land, e.g. worms aerate the soil.
3. Use minimum effort for maximum effect.
4. Increased yields, i.e. that it should be possible to increase the yield of a permaculture system through improving cultivation methods.
5. Outputs become inputs.
6. Each function should be supported by many elements.
7. Each element performs several functions.
8. Relative location, i.e. each element within the system should be located in the place most beneficial for the whole system.

Robert Hart's forest garden in Shropshire is an often used exemplar, taking as it does the interconnectedness of the forest and redesigning it to create an 'edible ecosystem'. Whitefield describes the elements, chosen carefully to maximise their interconnectedness: a canopy of fruit trees, a lower

Figure 22.1

Figure 22.1, 22.2 and 22.3 *Robert Hart's permaculture garden, an 'edible ecosystem'.*

Figure 22.2

Figure 22.3

layer of dwarf fruit trees and nut bushes, a shrub layer of soft fruit, a layer of perennial herbs and vegetables at ground level, plus root vegetables and climbers (Whitefield, 1997).

As Whitefield observes, each layer comes into leaf at different times of the year – the herb layer in early spring, followed by the shrubs, and lastly the trees – so that, through careful selection and placement of the elements, the greatest possible use is made of the available natural resources. The same can be said of the soil resource, since the plants' roots feed at different depths and each has a slightly different nutrient requirement. The cyclic nature of the ecosystem (principle 6, above) is also reflected, with some plants needing specific nutrients for growth, and some returning those same nutrients to the soil (principle 5, above). The system is also relatively self-maintaining, as it uses perennial plants (principle 3, above).

Another often cited exemplar is the permaculture chicken. Whilst there may be some animal rights issues here, it is worth noting that permaculture animals are well looked after and (by definition) allowed to live out their natural desires where at all possible – certainly to a much greater degree than the conventional barn or battery methods of 'husbandry'. In the

Figure 22.4 *Permaculture chickens.*

case studies I have researched, chickens and hens featured prominently. They could lay eggs for food use; their droppings could be used as fertiliser (principle 5, above); and they could be useful for controlling pests – they have been referred to as a 'free range slug patrol' (Sherriff, 1999). If a chicken house is connected to the side of a greenhouse then, as Whitefield (1997) enthuses: 'the greenhouse is kept warm by the heat of the chickens; the chickens are kept warm in the winter by the sun coming through to the greenhouse; and the carbon dioxide exhaled by the chickens fuels plant growth in the greenhouse.' – an excellent example of principles 5, 7 and 8, above.

So permaculture is distinguished partly by its use of nature as a model. Another characteristic is its ethical rootedness in sustainability. Robert Hart saw his gardening in the context of a sustainable vegan lifestyle, feeding himself almost entirely from the forest garden and some other vegetable growing.

Whitefield (1997) describes the three ethical foundations of permaculture as: earth care, that we must look after the earth in order to look after ourselves; people care, that sustainability should not be achieved at the cost of our freedom and our quality of life; and fair shares, which is about recognising the limits of the earth. These are clearly very similar to the definition of sustainability espoused at the 1980 Rio Conference, that development should 'meet the needs of the present, without compromising the ability of future generations to meet their own needs.'

Now we have arrived at a much more sophisticated definition:

> 'permaculture is essentially an approach to designing whole systems, through the maximisation of the interconnectedness of elements, which has an ethical foundation in sustainability and a scientific basis in ecology.'

One thing to note is that permaculture is different from organic agriculture. By organic agriculture I refer to produce such as that certified under European Union standards and policed by organisations such as the Soil Association. This is basically agriculture without pesticides and fertilisers, without genetically modified crops, and with compassionate animal husbandry.

Permaculture will often look like organics, and the end result of a permaculture design may qualify for organic certification, but there a number of important differences. Organics is a production method: permaculture is an approach to design and production.

Permaculture places more emphasis on cycling energy and resources locally; it places greater emphasis on the maximisation of interconnectedness; it is creative rather than regulatory; it

Figure 22.4

Figure 22.5 *Recycling plastic bottles as cold frames.*
Figure 22.6 *A typical small cold frame.*

emphasises the use of perennials; self-regulatory systems are encouraged; and community trading structures take a clear priority over global trading. The latter is particularly striking. Whilst organisations such as the Soil Association should be congratulated for campaigning for more farmers' markets and other local trading initiatives, you can still buy cabbages classed as organic that have been flown hundreds of miles for each of the processing, distribution and sale stages. This is in marked contrast to, for example, Hardy's Field in Lincolnshire, which has sold carrots through a Local Exchange Trading Scheme (Sherriff, 1999). The ecological bill of the global food system has to take account of the transport networks needed to facilitate the increasing transit of food and food-related products as well as the significant greenhouse gas emissions from the transport itself. Fruit and vegetables now form the largest category of air freight by weight (Friends of the Earth, 2001), and much of this could be grown more locally.

Having described permaculture, how can it inform urban agriculture? Using our definition of permaculture, we can make this question more precise: why should a group or person involved in the growing of food in urban areas equip themselves with a methodology that involves the conscious design of systems based on the principles espoused by Mollison?

My first observation from my research, is that this approach has a track record of producing a large amount and variety of food from a small area. If a city is trying to move towards self-sufficiency, monocultural cropping becomes less and less relevant. Permaculture provides an approach to making our plots more and more diverse. One of the case studies for my research, a London household, reported that it was able to get basic vegetables and greens all year round and potatoes from July until December, all from their permaculture garden. Additionally, they

Figure 22.5

Figure 22.6

made summer fruits into preserves for use during the winter (Sherriff, 1999). Robert Hart's example described above is another case of the significant way that permaculture can contribute towards self-sufficiency.

Where self-sufficiency is not the aim, permacultural trade systems extend the scope of the project from the individual to the community. The Local Exchange Trading Scheme (LETS) does for the community, what the permaculture garden does for plants: it maximises the interconnectedness between disparate elements, in this case, the needs and abilities of people in a community. Some members offer skills or products that others require, and they in turn offer another skill or product.

Vegetable box schemes are another way to distribute produce. Customers receive a box of mixed vegetables and fruit, with contents depending on the season, so the output of a permaculture plot or garden is perfectly suited. Farmers' markets are ideal for trading permaculture produce: gone is the need, experienced when dealing with supermarkets, to grow a certain amount of a certain crop. One of the beauties of the simplicity of permaculture is that when its principles are applied to food production and to trade, the two activities become symbiotic: the diversity of permaculture food suits local schemes, and the financial security of the trading methods enables a greater variety of crops to be grown and creative risks to be taken.

Permaculture food growing is never seen in isolation. Becon Tree Organic Growers, for example, develop the local economy through LETS, work with a local university, and with colleges and conservation groups (Sherriff, 1999). They are able to fulfill the criteria of Local Agenda 21 through: reusing and recycling resources, saving energy, cultivating local land, monitoring the local environment, green building and planning, community development and education, and developing the local economy.

Permaculture, by definition, seeks where possible to utilise resources frugally. This is important in urban agriculture for two reasons: it enhances its sustainability; and it makes it inexpensive to practice. The latter is particularly important where urban agriculture is proposed as a way to regenerated deprived areas. It is common for resources to be recycled. At Naturewise in London, old vehicle tyres were used to form a 'tyre garden' and in another London example old glass bottles were ingeniously used as edging between the lawn and the borders both to stop weeds spreading and to use as miniature greenhouses in which to germinate seeds. At Beacon Tree Organic Growers:

'... the pathways are made from heavy-duty rubber conveyor belts from a local gravel pit that are wide and strong enough to take an electric wheelchair and are covered with cockleshells form Leigh-on-Sea.' (Warwick, 2001)

From butts to collect rainwater, to composting toilets that turn human excrement into usable fertiliser, the efficient use and reuse of resources is central to permaculture's scientific and ethical foundations.

Another important aspect of permaculture is its avoidance of artificial fertilisers and pesticides, preferring instead to maintain soil health through a number of holistic techniques including polycultural planting and green mulching, and to deter pests through biodiverse planting and the encouragement of predators to frequent the ecosystem. Legumes, such as clover, can provide a crop with nitrogen, for example, and a biodiverse garden confuses pests and is therefore less vulnerable than a monoculture.

Overuse of artificial chemicals in farming is a significant environmental problem. The cost of removing agricultural pesticide residues from drinking water has been estimated at £119.60 million a year, and nitrous oxide from fertiliser use is a contributor to air pollution (Friends of the Earth, 2001). When thinking on the local and urban scale, water and air pollution becomes a more immediate threat to both producer and consumer than it does where food is produced and traded globally.

Figure 22.7 *The edge of a forest garden.*
Figure 22.8 *Grasses: permaculture and diversity.*

Figure 22.7

Figure 22.8

There is a growing body of evidence in support of concerns about artificial chemical use on health grounds. 45 pesticides, for example, are known or suspected hormone disruptors, potentially affecting the reproduction of both humans and wildlife. A recent study in Denmark found that women with higher than average levels of pesticides, such as dieldrin, in their bloodstream have double the risk of breast cancer (Sustain, 2001).

Conversely, produce grown without pesticides and fertilisers and in an organic regime are higher in nutritional value than those grown conventionally, due largely to the health of the soil. An American study found significantly higher levels of minerals in the former. There was reportedly 63 per cent more calcium, 73 per cent more iron, 125 per cent more potassium and 60 per cent more zinc (Sustain, 2001). Permaculturally produced food can be assumed to have similarly high standards since its attitudes to chemical use and the maintenance of soil health are comparable. It follows that any urban agriculture seeking to supply communities with a healthy alternative to the conventional would do well to follow the chemical-free and soil building principles of organic agriculture. As discussed above, permaculture provides an approach to creatively apply these principles at the local level.

Permaculture, then, is valuable as an approach to urban food growing and provides a useful methodology for food growing and local trading. This chapter has been able to give only a very brief description of permaculture. The references list contains many excellent sources of further information.

REFERENCES AND FURTHER READING

FoE (2001). *Get real about food and farming.* Friends of the Earth.

Molllison, B. (1991). *Introduction to Permaculture.* Tagari.

Mollinson, B. (1992). *Permaculture a designer's manual.* Tagari.

Pepper, D. (1996). *Modern Environmentalism.* Routledge.

Rob, H. (2000). The Food Producing Neighbourhood. In *Sustainable Communities: The Potential for Eco-Neighbourhoods* (H. Barton, ed) Earthscan.

Sherriff, G. (1998). Edible Ecosystems. In *Sustainable Agriculture: A Study of Permaculture in Britain*, Keele University, Staffordshire, available at www.edibleecosystems.care4free.net

Sustain (2001). *Organic food and farming: myth and reality*. Sustain.

Warwick, H. (2001). Urban renaissance. In *Permaculture Magazine* Issue 30, Winter 2001.

Whitefield, P. (1997). *Permaculture in a Nutshell*. Permanent Publications.

23

**UTILITARIAN DREAMS:
EXAMPLES FROM OTHER COUNTRIES**

André Viljoen

If Cuba can be considered a laboratory for the wide scale introduction of urban agriculture, it is not alone in planning for urban agriculture. Examples can be found in Asia, Africa and Europe. Although the conditions in each location are different, a number of common benefits can be identified.

DELFT, THE NETHERLANDS

The city of Delft in the Netherlands provides an interesting example of planning legislation being adapted to accommodate peri-urban agriculture. (Deelstra et al., 2001)

Urban agriculture in Upper Bieslandse Polder
Urban areas
Road
Highway
Railroad
Water

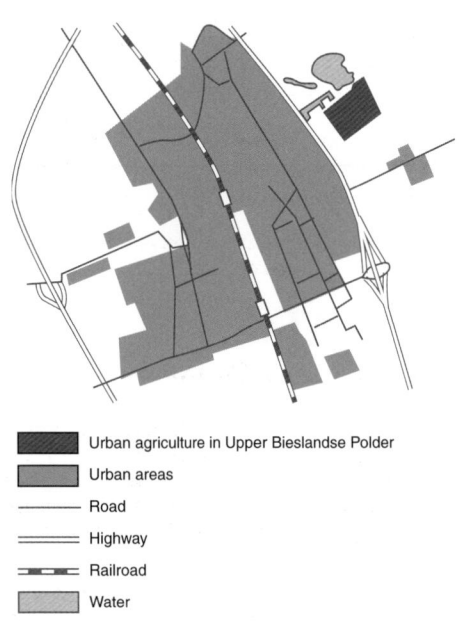

Figure 23.1

The Upper Bieslandse Polder has an area of thirty-five hectares and lies on the eastern edge of Delft. The land had been rented to farmers on short leases pending development. Collaboration between farmers, environmentalists and planners resulted in the city authorities granting a twelve-year lease to a farmer committed to organic dairy farming. This project demonstrates a number of benefits arising from urban agriculture. The organic farm includes designated areas for wildlife habitat, which have been located on the perimeter of the farm providing a transition between public and private land. Footpaths, cycle ways and bridal paths have been included in the plan, further intensifying the farmland use by including recreational areas for Delft's inhabitants. Water meadows, marshy woodland and reed beds provide a degree of ecological water management. We could call this kind of environmental development, horizontal intensification. It contrasts sharply with the notion of vertical intensification currently being explored by many Dutch architects. Vertical intensification proposes a stacking of different activities and ecological systems within a multi-storey structure. Vertical intensification proposes highly engineered artificial landscapes. While horizontal intensification as found in the Upper Bieslandse Polder is also artificial, in that the land is managed, the inputs in terms of energy and materials are far less, and the outputs are almost entirely environmentally benign.

Although the proposal may be imperfect, in that it relates to a single example of land developing horizontal intensification, and it is impermanent (twelve years), it does indicate how policies are beginning to change in Europe.

KATHMANDU VALLEY, NEPAL

Some similarities can be identified between policies in Delft and development guidelines introduced in

Figure 23.1 *Upper Bieslandse Polder, Delft, Netherlands.*

UTILITARIAN DREAMS

are extensive. We would estimate that if a similar proportion of land was allocated in European cities, they would be close to self-sufficient in fruit and vegetables. But one imagines the opposition to such proposals if powerful developers had plans for particular pieces of land. This highlights the need for planning for urban agriculture at an early stage in the planning process, preferably before land has been sold to individuals for the purpose of house building.

The Madhyapur Thimi Municipality has identified a number of significant benefits resulting from the inclusion of urban agriculture reserves.

By defining urban agriculture reserves, financial resources can be targeted more efficiently, as areas of land are identified which do not require the installation of complex supporting infrastructure. This allows the concentration of resources within areas identified for urban development. Resources can be used to promote denser urbanisation and the efficient use of land. Furthermore, potential problems, such as processing biodegradable waste and sewage, can be turned into solutions, i.e. if this waste is used as an agricultural resource, for example as compost. By having a specific policy supporting urban agriculture, the release of any agricultural land for development can be regulated and controlled.

- Urban agriculture reserve areas
- Urban development areas
- Buildings
- Roads

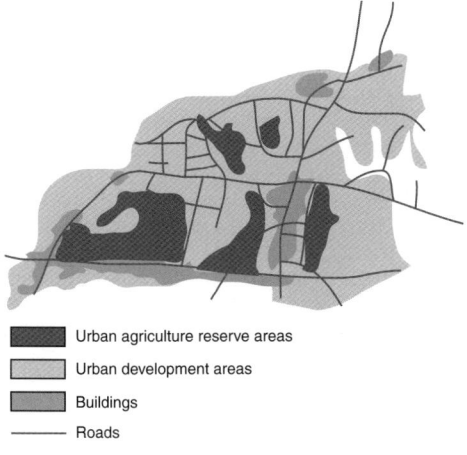

Figure 23.2

Nepal's Kathmandu Valley. An extensive planning exercise undertaken for the Madhyapur Thimi Municipality, located in the Kathmandu Valley, has defined a number of urban agricultural reserve zones, which have been incorporated into the municipality's development plans (Weise and Boyd, 2001). The reserves are not necessarily permanent as the municipal plan allows for the incremental release of land for commercial development. In this respect proposals are similar to those in Delft, namely that planners have yet to be convinced that urban agriculture can be considered as beneficial as conventional urban development. Agreement to incorporate the urban agriculture reserves into the municipal plan was not without opposition, in the main from middle class individuals who had intended to build detached houses on small plots. A portion of land has been set aside for housing, but the urban agriculture reserves in Madhyapur Thimi Municipality

GABORONE, BOTSWANA

In Gaborone, capital city of Botswana, a number of integrated urban agriculture projects have been instigated. In at least one case, innovative planning has made a connection between waste from sewage works and urban agriculture (Cavric and Mosha, 2001).

Gaborone has undergone dispersed suburban expansion since independence in 1966. The city

Figure 23.2 *Madhyapur Thimi Municipality, Kathmandu Valley, Nepal.*

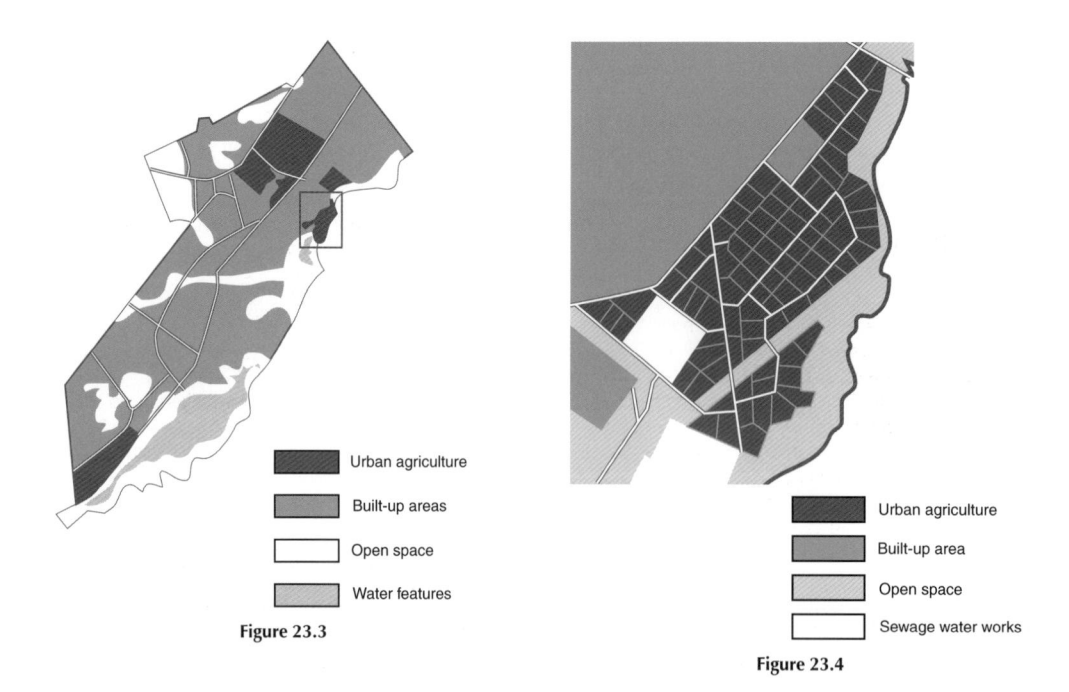

Figure 23.3

	Urban agriculture
	Built-up areas
	Open space
	Water features

Figure 23.4

	Urban agriculture
	Built-up area
	Open space
	Sewage water works

has always been reliant on food imported from beyond its boundaries (Mosha and Cavric, 1999). In the Glenn Valley area of Gaborone several parcels of land have been identified as sites for urban agriculture. It is estimated that these should make the city self-sufficient in food production. One of these sites is of particular interest as it is adjacent to a sewage works and bounded by the Notwane River. Wastewater from the works can be used to irrigate crops, though if industrial contaminants are present wastewater should be reserved for non-edible crops such as fiber or wood. In the Glenn Valley area individual farmer's plots vary in size between one and a half and four hectares and are laid out back to back in order to minimise the number of access roads required.

The plots adjacent to the sewage works are likely to be used for cut flowers and ornamental plants, similar to the floral organoponics found in Cuba. We can use this example to imagine how such an area, including the urban agriculture plots, could become a new kind of urban park. As designed, the Glenn Valley urban agriculture site is laid out as a matrix of rectangular plots. If these plots are separated to allow for continuous paths between them, a network of accessible public routes could overlay the urban agriculture fields, creating what we

Figure 23.3 *Glenn Valley, Gaborone, Botswana.*
Figure 23.4 *Glenn Valley (detail), Gaborone, Botswana.*

UTILITARIAN DREAMS

term 'LeisurEscapes'. This is another example of horizontal intensification, in this case intensifying a productive landscape by making a connection between the river and the city, while providing new spaces for play and relaxation.

ANOTHER MODEL

In the previous examples specific parcels of land have been set aside for urban agriculture. An alternative strategy has been developed in Tanzania and Bulgaria, where a less specific categorisation or zoning is applied. In each case it has been proposed that as one moves away from the urban centre, the potential for including urban agriculture increases, as more open space becomes available. Thus particular parcels of land are not zoned for food growing, but the extent of urban agriculture acceptable in different areas is defined.

Dar es Salaam, Tanzania

In Dar es Salaam, Tanzania, the city's Strategic Urban Development Plan now accepts urban agriculture as a legitimate land use, where before it was only tolerated as a transitional land use (Kitilla and Mlambo, 2001). Kitilla and Mlambo refer to the rapid increase of informal urban agriculture in the city, for example between 1985 and 1993 the city witnessed a thirty-fold increase in the number of goats. So planning policy has had to react to a situation which is occurring anyway. It is estimated that 30 per cent of the food consumed in Dar es Salaam is grown within the city boundary.

As a consequence of supporting urban agriculture, the city authorities promote the vertical expansion of buildings, where this can free up urban land for food growing. In Dar es Salaam, the availability of low cost transport has been recognised as essential for the transportation of locally grown food. Here again we can imagine how interconnected parcels of productive landscape can generate horizontal intensification, and by providing routes for the distribution of goods by, for example, bike, as in Cuba, also provide an infrastructure for local commuting and leisure.

Trojan, Bulgaria

The town of Trojan in Bulgaria provides an example of places where urban food growing occurs in

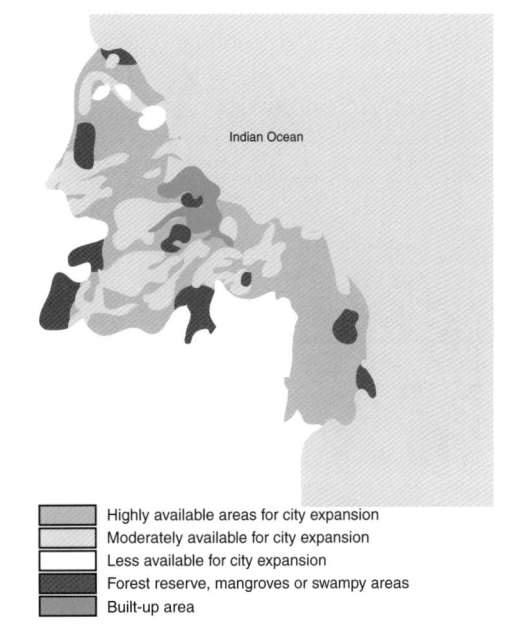

Figure 23.5

Highly available areas for city expansion
Moderately available for city expansion
Less available for city expansion
Forest reserve, mangroves or swampy areas
Built-up area

Figure 23.5 *Dar es Salaam, Tanzania.*

Figure 23.6

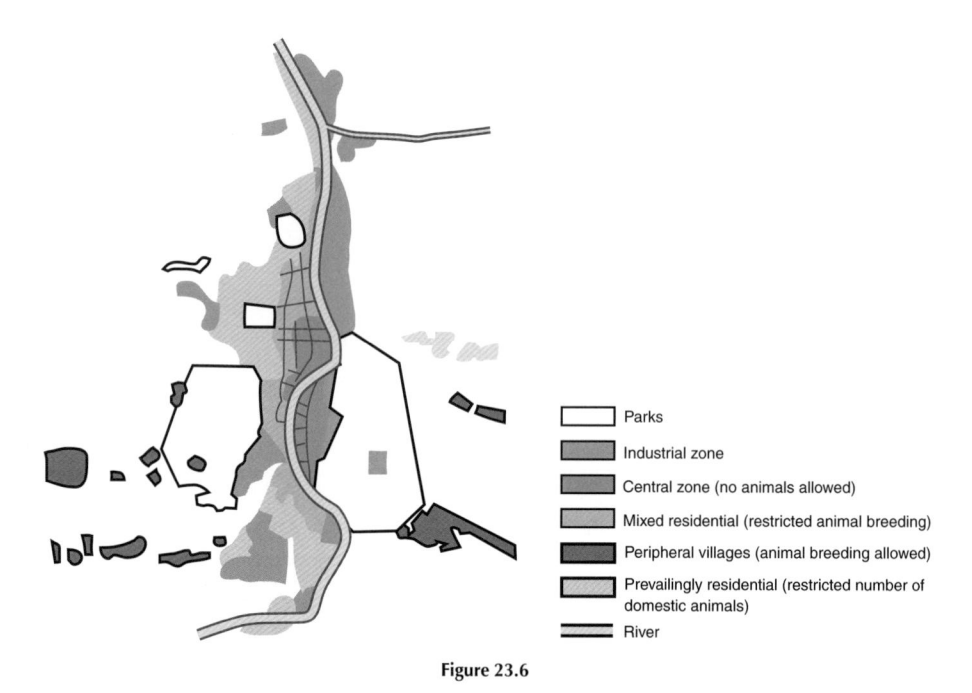

- Parks
- Industrial zone
- Central zone (no animals allowed)
- Mixed residential (restricted animal breeding)
- Peripheral villages (animal breeding allowed)
- Prevailingly residential (restricted number of domestic animals)
- River

residential areas and is consumed by the growers, rather than being sold. Although urban agriculture has not been considered in urban planning, the fact that it exists and makes a significant contribution for a section of the population has resulted in the development of 'Experimental Rules' for urban agriculture (Yoveva and Mishev, 2001).

Planners instigated a participatory process, with local farmers and gardeners, to investigate the development of planning policy for urban agriculture. An action plan for the promotion of peri-urban agriculture has been adopted, which commits the municipality to co-ordinating support structures for the implementation of agricultural projects.

In the initial stages most gardeners have expressed an interest in self-sufficiency, rather than developing commercial market gardens, however the planning team have identified the development of agro-tourism and landscape conservation as long-term goals for their productive urban landscapes.

The example of Trojan raises significant questions in relation to suburbia and sprawl. Yoveva and Mishev observe that urban agriculture in Bulgaria,

Figure 23.6 *Trojan, Bulgaria.*

as practised by individuals for self-consumption, is largely provoked by financial hardship as the country undergoes a process of transition to an open market. From this we can make a reasonable deduction that the majority of the population do not wish to grow their own food, and hence our contention that sustainable cities should make provision for productive landscapes which contain commercially viable mini or supermarket gardens (Viljoen and Tardiveau, 1998). But, by rejecting the notion of individual self-sufficiency, we should not lose site of the opportunities within existing suburbs for extensive food growing. Studies from Auckland in New Zealand have indicated the significance of garden vegetable plots within suburbs and the trend for these to become under-utilised in recent years (Ho, 2000). This extensive area of under-utilised land suggests ways in which portions of private gardens could be connected into a network of linear fields, farmed professionally by gardeners who would supply produce to home owners.

Readers who would like more information about these and other initiatives are directed to *The Urban Agriculture Magazine*, Number 4 (2001) published by The Resource Centre for Urban Agriculture and Forestry, RUAF. Copies may be downloaded from their website at www.ruaf.org.

REFERENCES

Cavric, B. I. and Mosha, A. C. (2001). Incorporating Urban Agriculture in Gaborone City Planning. *Urban Agriculture Magazine*, Number 4, 25–27.

Deelstra, T., Boyd, D. and van den Biggelaar, M. (2001). Multifunctional Land Use: An Opportunity for Promoting Urban Agriculture in Europe. *Urban Agriculture Magazine*, Number 4, 33–35.

Ho, S. (2000). Food production in Cities. *Proceedings of Shaping the Sustainable Millennium*, University of Brisbane. (available at www.sdrc.auckland.ac.nz/confpprs.htm)

Kitilla, M. D. and Mlambo, A. (2001). Integration of agriculture in city development in Dar es Salaam. *Urban Agriculture Magazine*, Number 4, 22–24.

Mosha, A. C. and Cavric, B. (1999). *The practice of UA in Gaborone*. Department of Environmental Science, University of Botswana research project.

Weise, K. and Boyd, I. (2001). Urban Agriculture Support Programme for Madhyapur Thimi Municipality, Nepal. *Urban Agriculture Magazine*, Number 4, 33–35.

Viljoen, A. and Tardiveau, A. (1998). Sustainable Cities and Landscape Patterns. *Proceedings of PLEA 98 Conference*, Lisbon, pp. 49–52.

Yoveva, A. and Mishev, P. (2001). The Case of Trojan Using Urban Agriculture for Sustainable City Planning in Bulgaria. *Urban Agriculture Magazine*, Number 4, 14–16.

PART FIVE

CARROT AND CITY: PRACTICAL VISIONING

24

NEW SPACE FOR OLD CITIES: VISION FOR LANDSCAPE

Katrin Bohn and André Viljoen

SIZE

Figure 24.1 contrasts the size of a nineteenth century London park with proposals for a modest CPUL intervention. It is not the size of an individual urban agriculture site that determines its success as a Productive Urban Landscape. Size will be significant in determining the yield and hence the environmental impact of urban agriculture sites (see Chapter 3), but it is not critical in relation to the qualities brought to a city. We need to distinguish between individual plot size and the extensiveness and interconnectivity of productive urban landscapes. It is interconnectivity that will lead to continuous landscapes, which can ultimately generate a new Ecological Infrastructure in urban environments. Productive urban landscapes may consist of many small fields covering an extensive area, or of isolated patches of horticulture set far apart, or of large individual fields. Fingers of productive landscape may link, like bridges, associated but physically isolated activities and areas of the city. Any one piece of land supporting urban agriculture may vary in size from several square metres in area to several hectares. Implementing productive urban landscapes may start at a small scale, but the goal is to develop CPULs. At its greatest extent, a network of green spaces would overlay and run through the urban fabric, with agricultural fields placed within a continuous landscape.

The authors, have tested the idea of CPULs in a number of design studies. LeisureESCAPE, is a study for one of several strands of continuous landscape, which could be introduced into London (see Plate 6). The study examines an area running south of the Tate Modern art gallery, in the city centre, and continuing for about twenty kilometres to the edge of Greater London, where the continuous landscape disperse into the countryside. A detailed image of part of this scheme shows the relatively modest number of roads required to be given over to growing to create an extensive network of productive urban landscapes. (see: colour plate 8)

Productive landscapes and urban agriculture can exist at a much more modest scale, when located within the boundary of a particular site. Figure 24.2 shows a modest proposal for such a small-scale intervention, sited between two blocks of live work apartments. This landscape would be like a shared square, like a small version of the crescents found in Bath, or expanded versions of Berlin courtyard apartments. These micro fields would be used daily by residents living in adjacent buildings and thus planting would be somewhat more ornamental in character, soft fruits and berries, for breakfasts of strawberries and pears ripe from trees outside a bedroom window.

It is probable that in many instances the courtyard or square will provide a model for Productive Urban Landscapes. In Figure 24.3 we can see plans for a riverside development, which places a number of urban agriculture fields between a riverbank promenade and terraces of apartments and houses. Here the situation closely resembles examples found in Cuba, as seen in Plates 4 and 5.

Productive Urban Landscapes need not be horizontal. Certain places in a city may be appropriate for vertical landscapes, where one or more layers of vegetation are placed vertically against a building's facade. These second skins provide ecological and environmental benefits. They also provide connections between ground-based fields and cultivated aerial platforms, offering a further degree of ecological intensification, this time an expansion of ground area. Vertical landscapes present a particular form of ecological intensification, by artificially increasing the carrying capacity of a piece of ground.

NEW SPACE FOR OLD CITIES: VISION FOR LANDSCAPE

Figure 24.1

This intensification requires very specific design and maintenance once implemented. Dealing with the difficulties associated with gardening in shallow soils, at heights where exposure to drying, cold, and overshadowing is frequently extreme, will limit the application of vertical landscapes to particular sites and buildings. The expert management required for vertical landscapes suggests that they

Figure 24.1 *Victoria Park CPUL: A modest CPUL adjacent to Victoria Park in North London. Interventions like these retain the character of larger CPULs and may, in the future, be extended.*

CARROT AND CITY: PRACTICAL VISIONING

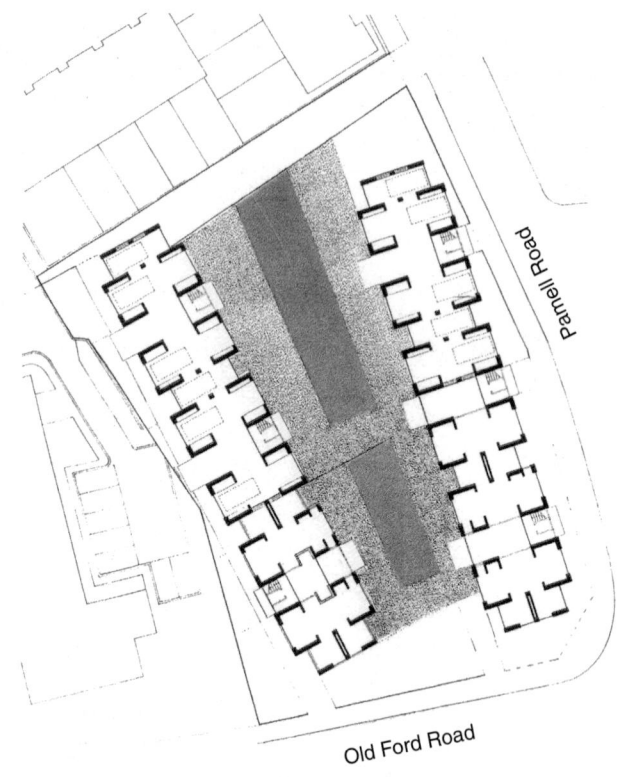

Figure 24.2

are best developed at a large scale on a single building or on a cluster of smaller buildings, thus facilitating maintenance by a team of dedicated farmer/gardeners. Plate 15 and Figure 24.14 show one example of the integration of vertical fields within a dense development in the city of London.

SENSE OF OPENNESS

It is reasonable to expect that during the first stages of implementation of CPULs, a series of small interventions will, be made, eventually leading to an extensive network of connected spaces. Such

Figure 24.2 *Victoria Park CPUL: Small pieces of land, such as this one between two live/work units, can accommodate micro fields that are productive but domestic in character.*

NEW SPACE FOR OLD CITIES: VISION FOR LANDSCAPE

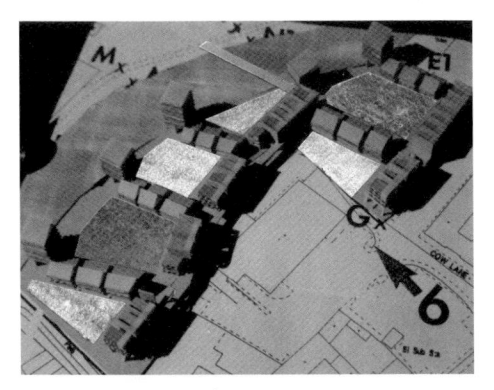

Figure 24.3

an approach will, over time, create a sense of openness within an otherwise uniformly built and occupied environment. Figure 24.4 shows the outcome of this kind of strategy, whereby over time disused and abandoned sites become activated and used in an environmentally and socially productive manner. Here, productive landscapes occupy interstitial spaces found in the gaps between a predetermined urban grid. Meaning and a sense of openness are introduced to the site by delineating expansive views and long vistas, to what would otherwise remain a series of isolated, disconnected and largely disused patches of land.

CPULs or patches of productive urban landscapes can provide space for wondering eyes, and receptive ears. Spaces to view fields overlooking a horizon can be created within modest developments. Figures 24.5 and 24.6 show urban agriculture fields in Newark, set between housing terraces. Windows from within, and the paths alongside these houses, offer views across fields and sky. These proposals

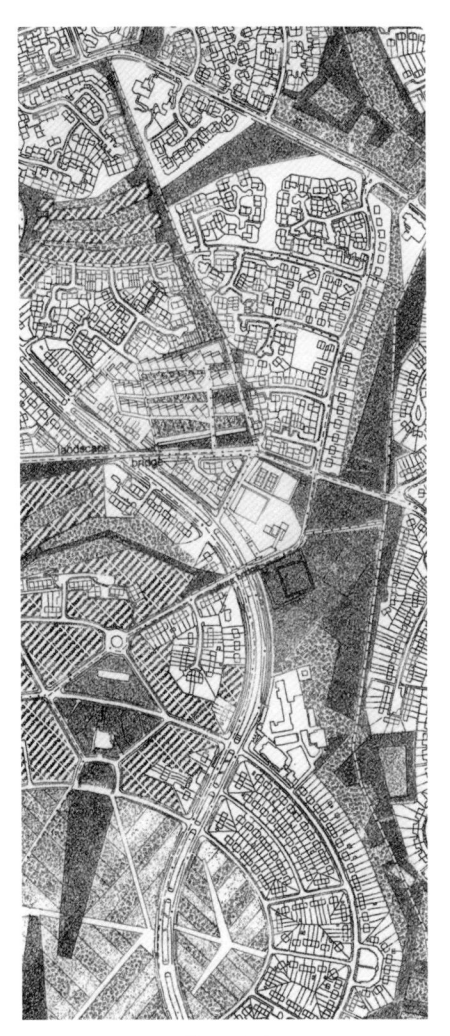

Figure 24.4

Figure 24.3 *Newark CPUL: Urban agriculture fields are situated like courtyards between housing and a river.*
Figure 24.4 *ElastiCity CPUL, Sheffield: In a strategy for developing CPULs over time, disused and abandoned sites become activated and reused in an environmentally and socially productive manner, creating a comprehensible and useful environment.*

CARROT AND CITY: PRACTICAL VISIONING

Figure 24.5

Figure 24.6

Figure 24.7

embody characteristics described in the Edge Atlas (see Chapter 17-Edge atlas).

The scale at which these conditions are experienced can be varied according to the degree to which occupation is in the private or public realm. In our proposals for Sheffield, (see Figure 24.7) the insertion of a courtyard into a dwelling allows eyes and occupants to move between interior and exterior within the domestic realm. Through this courtyard, access is provided to a linear landscape, accommodating a mix of small and large patches of urban agriculture and recreational spaces. The height and size of boundaries guide eye and body through and across the landscape. Plate 13 shows how the eye is drawn from within to without, a condition that reoccurs at a variety of scales.

LOCAL INTERACTIONS

Benefits occur no matter what the scale of intervention. Modest linear fields (Figure 24.1), can provide space for paths, which connect private and public spaces. Making their adjacency visible encourages movement between the two. The paths and connecting landscape can be thought of as interventions

Figure 24.5 *Newark CPUL: Houses, paths and fields.*
Figure 24.6 *Newark CPUL: Houses look out over fields, towards a river beyond.*
Figure 24.7 *ElastiCity CPUL, Sheffield: Garden/courtyards mediate between the private realm of the house and the public realm of a CPUL. A two-storey core home can be enlarged by building single-storey additions in the garden/courtyard.*

244

Figure 24.8

which mark and reveal. Routes to shops become adjacent to places where food is grown. Each walk amongst the crops heightens the experience of seasonality, and speeds up time because of the compact space within which nature is experienced. Time is intensified – more 'nature' for your time. No need to pack your bags to visit the countryside.

Figure 24.8 indicates improved access and interaction at its most elementary. Here houses are elevated above an urban agriculture field, their exact height determined by a requirement to provide access, for cultivators, to fields and to maximise solar access to the fields. At Newark ramps from the front door connect the houses to a river at one side and the city centre at the other. Vistas are created over the urban agriculture fields and the river, and houses float in a pastoral landscape.

Space promoting and provoking local interactions can be thought of as the space of encounters. (see Chapter 2). Imagine setting out into a continuous landscape, such as that proposed for Sheffield (see Figures 24.4 and 24.7). An individual will pass people engaged in different kinds of growing and cultivation. A pedestrian may come across farmers using their land commercially, while also passing allotment holders engaged in activities that blend the social and productive. These encounters occur within the boundaries of a continuous productive urban landscape. But the site within which local interactions occur, extends beyond the boundary of individual fields or landscapes. By setting up connections, continuous landscapes expand their sphere of influence to include adjacent enterprises, recreational areas and institutions. The act of making and marking a connection promotes and engenders physical contact. In certain instances, fine bridges can span roads or rivers, breaking what are otherwise insurmountable barriers. Figure 24.9 demonstrates the strength of such a landscape bridge, proposed as a lightweight structure, providing spatial continuity across a road which, otherwise, destroys any sense of adjacency.

URBAN NATURE

Natural landscapes, those that are wild and unmanaged, exist in the city with a diminished significance,

Figure 24.8 *Newark CPUL: Elevated terraced houses step down towards an adjacent riverbank. The broken-up terrace, sloping Westwards, minimizes overshadowing of urban agriculture fields and provides continuity of surface and access between fields.*

CARROT AND CITY: PRACTICAL VISIONING

Figure 24.9

both in terms of the number of these places and the recognition of their significance as a valuable ecological resource. Places often understood as natural, for example parks, are artificial, constructed to provide an idea or image of outside-town, of the rural, the pastoral. CPULs will be part of this constructed idea of the natural. Urban agriculture in particular, will represent a notion of the countryside, of rural life and, via this image, of 'nature'.

This is not to deny the power of the natural to occupy the urban, e.g. kestrels nesting in the city. But such urban nature is largely hidden, unseen and thus not registered. CPULs will trigger and support the possibilities for urban nature to establish itself and/or to expand. The adjacency of architecture, of urban life to agriculture and CPULs will reveal what is now hidden. Seasonality, as expressed by changing crop types and their ever changing appearance from sowing to harvest, by the exposure of the earth when it is laid fallow, by associated smells, sounds and views, will make the environment once more comprehensible. CPULs will intensify the connection occupants have with the living environment, without requiring the return to a subsistence economy, but equally reducing our destructive impact on the planet.

Figure 24.10

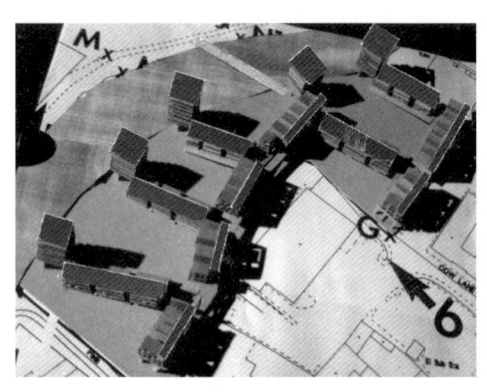

Figure 24.11

Planning strategies for brownfield or greenfield sites can include CPULs within architectural proposals. Figures 24.10, 24.11 and 24.12 show how fingers of dwellings run into urban fields, allowing

Figure 24.9 *ElastiCity CPUL, Sheffield: A landscape bridge provides physical and visual connections between CPULs either side of a main road.*
Figure 24.10 *Newark CPUL: Routes for pedestrians and cyclists highlighted.*
Figure 24.11 *Newark CPUL: Rooftop solar collectors highlighted.*

NEW SPACE FOR OLD CITIES: VISION FOR LANDSCAPE

continuity within spaces by giving simultaneous access to dwellings and agricultural land. The further layering of infrastructures and energy systems provides another, more subtle exposure to urban nature, one embodied in the concept of ecological intensification.

When density of population increases, more extreme measures are required, including vertical landscaping, mixed use occupancies and park-like public open space. Figures 24.13 and 24.14 show the relationship between vertical and horizontal urban agriculture fields. An undulating public park runs over perimeter accommodation, making an urban cliff edge overlooking horizontal fields of

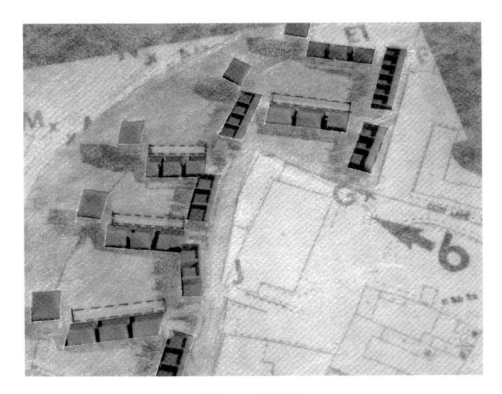

Figure 24.12

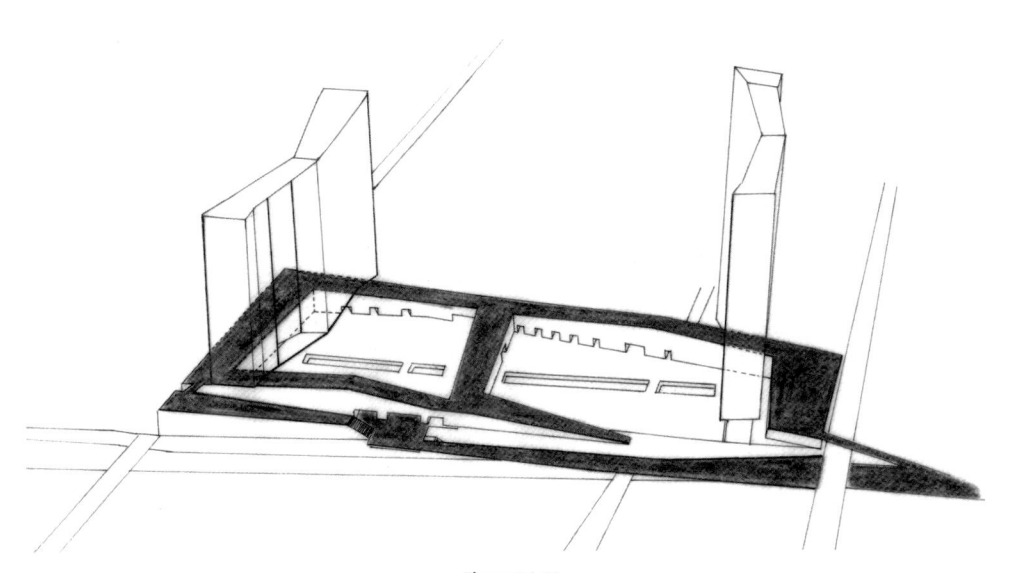

Figure 24.13

Figure 24.12 *Newark CPUL: Surfaces for collecting rainwater for domestic use highlighted. Underground storage tanks indicated by dashed lines.*

Figure 24.13 *Shoreditch CPUL: An undulating public park runs over perimeter accommodation and makes an urban cliff edge overlooking horizontal fields of urban agriculture. This park connects to existing open space which includes an adjacent city farm.*

CARROT AND CITY: PRACTICAL VISIONING

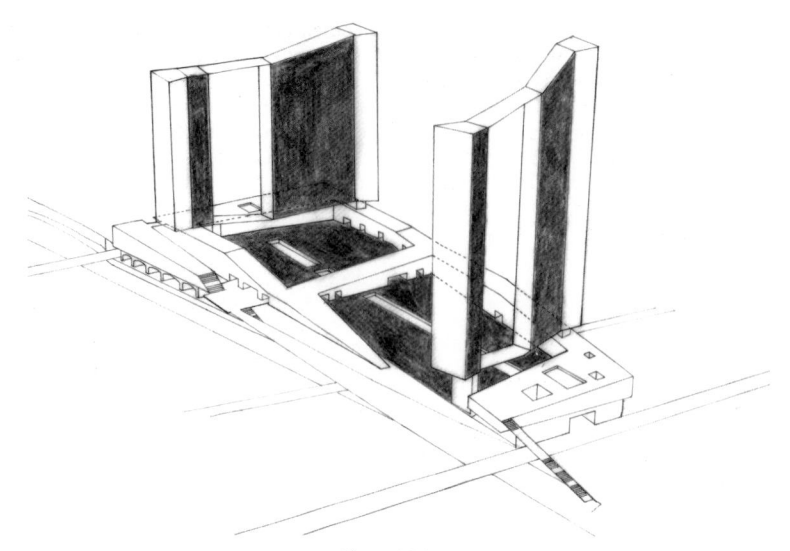

Figure 24.14

urban agriculture; the park, providing space for circulation and pleasure. People can picnic alongside an urban cliff edge, viewing across fields of urban agriculture.

In some buildings with a vertical landscape, a room will extend outwards from the building's facade creating a balcony, which cuts through the landscape curtain; here the occupant is placed in direct contact with a planted environment. In many cases the intimacy would be greater than that achieved in a rural location. Here a landscape is created in the sky, the planting creating a buffer between interior and exterior and framing a private space.

For the inhabitant of a building with a vertical landscape, the proximity of planting adjacent to a facade will provide opportunities for a direct experience of vegetation, in a context which usually emphasises dislocation from the ground and natural processes. A vertical landscape adjacent to a facade can be thought of as a thick green curtain hung in front of the building. Where windows occur, holes would be punched through this green curtain. By controlling the dimensions of the openings in the green facade, experiences of the planted landscape would be modified. By separating the vertical skin from the facade, it will provide solar shading to openings. By configuring openings which will not provide excessive obstruction to daylight and using openings to shade elevations facing direct solar gain, a variety of opening sizes will be generated, which will articulate facades in relation to their orientation. Thus, the developed landscape curtain will read like a fabric laid over the facade, a temporally changing surface

Figure 24.14 *Shoreditch CPUL: Productive fields in a dense development, including horizontal and vertical fields.*

(see Plate 15). It will have openings across the facade, Setting up spatial relationships between the building as a static solid object and the veiling second skin created by the landscape curtain. With the passing of seasons, the landscape curtain's surface will change in colour, density and opacity. It will be heard, producing a rustle in the wind. And pedestrians, walking past such a building, will experience the facade as an animated surface.

PERSISTENT VISUAL STIMULATION

Rivers and fields contribute to persistent visual stimulation, a characteristic of productive urban landscapes. In some cases direct relationships exist between spaces with different temporal rhythms, for example a river, with its frequent surface undulations reflecting a dappled light, and a gentle sound, seen against the slower rhythms of crop lifecycles.

The arrangement of furrows and beds for planting, which are found in typical urban agriculture sites and market gardens the world over, echo ancient patterns of spatial divisions, found in nature and agriculture (see Plate 16). The artist Tom Phillips, has speculated that primal marks found in the natural landscape, for example line, point, and hatch, as well as characteristics such as branching, forking, repetition, and variation, provide a basic visual grammar and syntax, from which ornament has been derived (Phillips, 2003). The same words used by Tom Phillips to describe the natural environment can be used to describe the conditions found in urban agriculture fields. Continuing this train of thought it can be argued that the appeal of agricultural landscapes derives from an ancient understanding of the marks they make in the landscape, which can be read as an elemental form of ornament. Thus, urban agriculture in particular can be read as a deeply understood form of ornament within the city. Through urban agriculture it is possible to experience urban ornament.

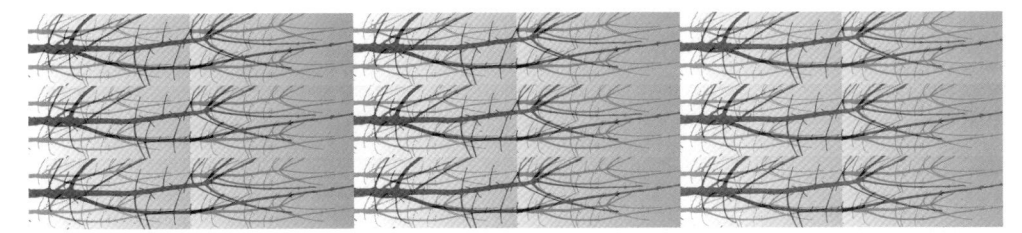

Figure 24.15

REFERENCES:

Philips, T. (2003). The Nature of Ornament: A summary Treatise. *Architectural Review*, Vol. CCXIII, No. 1274, 79–86.

Viljoen, A. (1997). The environmental impact of Energy Efficient Dwellings taking into account embodied energy and energy in use. In *European Directory of Sustainable and Energy Efficient Building 1997*. James and James (Science Publishers).

Viljoen, A., and Bohn, K. (2000). Urban Intensification and the Integration of Productive Landscape. July 2000. *Proceedings of the World Renewable Energy Congress VI, Part 1*, pp. 483–488, Pergamon.

LeisurESCAPE London

LeisurESCAPE touches
an **archaic desire** -
  the desire to move *in leisure*
  through open *space*.

It juxtaposes
human leisure desires
with proposals arising from the urge for an
independent sustainable urban future.

LeisurESCAPE enables
city dwellers to escape into the countryside
and country dwellers to escape into the city.

LeisurESCAPE is applicable to any
urban environment, but
most needed in large cities

LeisurESCAPE forms a
continuous landscape
running from outside London to the
Thames and to outside London again.

It works by **inter-connecting** existing
parcels of open land:
parks, playing fields, brownfield sites,
underused green spaces, public gardens,
large car parks....
through a **slim continuous landscape-**
LeisurESCAPE.

LeisurESCAPE is active and seasonal:
walking, talking, cycling, pushing (prams
and wheelchairs) scating...scating...,sitting,
laying, (sun-) (rain-) bathing,
reading...
jogging, hopping, sleeping...

This leisure landscape brings different
leisure activity areas **into proximity**
with each other and open space.

legend

continuous landscape discussed here

other continuous landscapes

continuous landscape dispersing into
countryside beyond the Greater
London Boundary

Thames Paths - partly continuous
existing path to one or both sides
of the River Thames

junction points

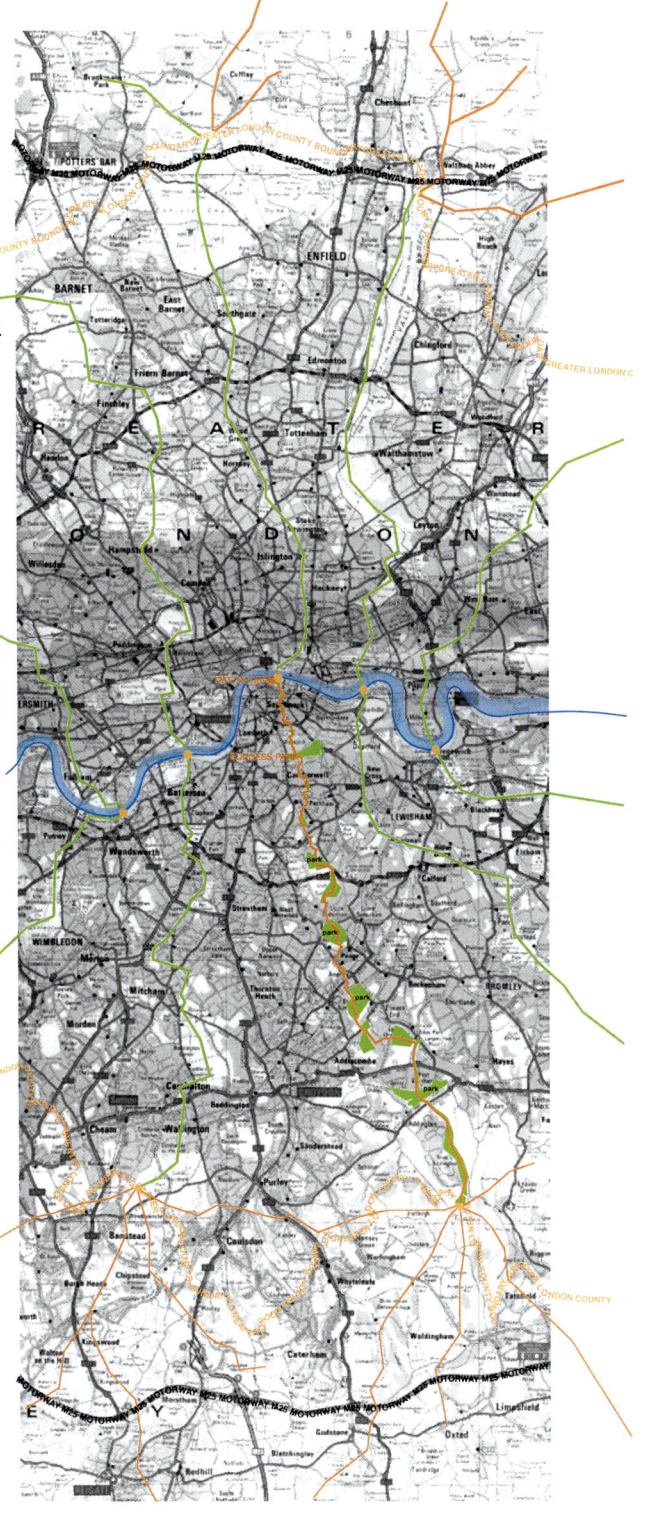

Plate 6

LeisurESCAPE London **Southwark** detail

LeisurESCAPE allows a **multitude of occupation**, both professional and leisure for all age groups, social levels, genders...

It caters especially for **population groups** which are often excluded from conventional leisure activities.

LeisurESCAPE is commercially and socially viable reinforcing the ecology and sustainability of the proposal.

The continuous landscape which accommodates LeisurESCAPE is laid out mainly over existing roads **based on the future vision of** reduced city car traffic.

Instead of the conventional usage of roads, **LeisurESCAPE** turns roads into a unique productive landscape **growing fruit and vegetables** for the city dwellers own consumption.

Agriculture fields in **LeisurESCAPE** are run both commercially and privately, thereby determining economic and social value.

LeisurESCAPE can provide new employment opportunities in its large areas of commercial agriculture or adjacent leisure facilities.
> Half of Southwark's population are pensioners. The number of lone parents is above the national average and rising.

LeisurESCAPE is **adaptable and slow** and creates opportunities for the growing number of pensioners, lone parents with toddlers, the disabled or unemployed.

Successful precedents are, for example:
Cuba (commercial - Organoponicpos)
Austria (leisure - Selbsternte)
or Germany (leisure - Schrebergarten).
Nevertheless, none of those combines explicitly commercial and leisure activities and not in a continuous landscape.

legend

■ continuous landscape

■ existing parks

■ underused open space

■ semi-buried existing large car parks, covered with productive landscapes

■ existing playgrounds and playing fields

■ small specific leisure buildings

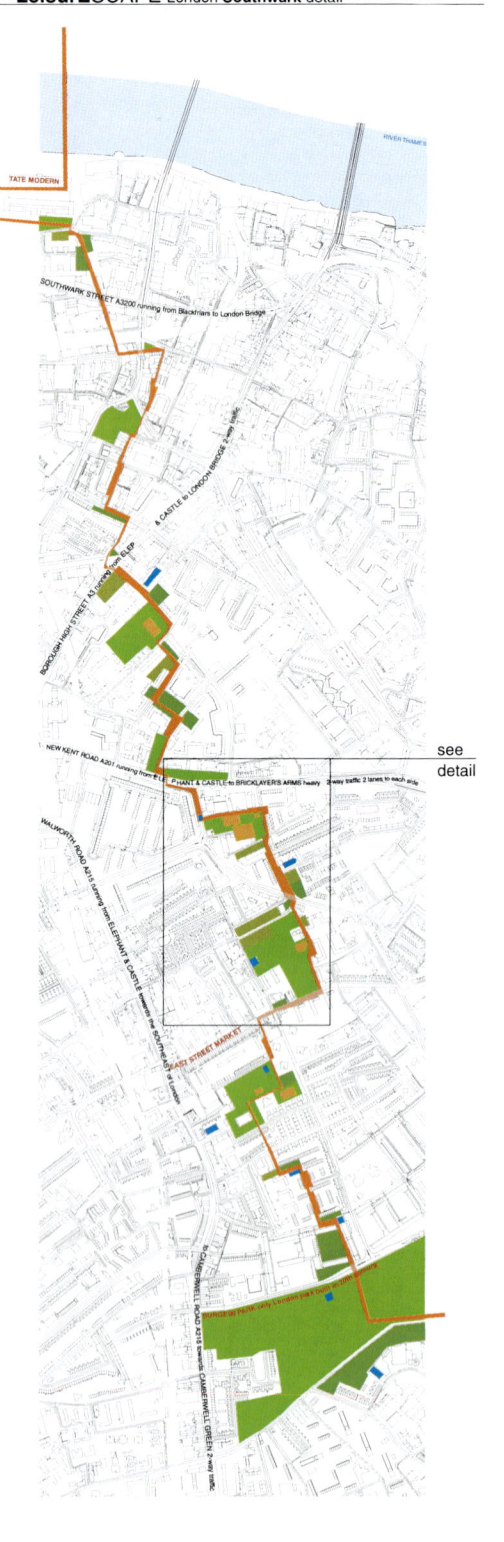

Plate 7

LeisurESCAPE London **Southwark** detail

Plate 8

LeisurESCAPE London **Orb Street** Southwark

Continuous landscape and productive landscape,
Orb Street Southwark, before and after.

The installation of **LeisurE**SCAPE:
horizontal, vertical and espaliered vegetation, playing fields and a covered car park

Plate 9 and Plate 10

LeisurESCAPE London **Munton Road** Southwark

CPUL infrastructure,
Munton Road, Southwark, before and after.

The installation of **LeisureE**scape:
footpaths, cycle networks, market gardens infrastructural intensification

Plate 11 and Plate 12

CPULs edge buildings

Manor Estate, Sheffield:
View from a dwelling into walled garden with windows opening onto a productive landscape beyond.

Victoria Park, London:
Transitions between interior, exterior, private and public, in an elevated apartment set between a rooftop garden and a productive landscape on the ground.

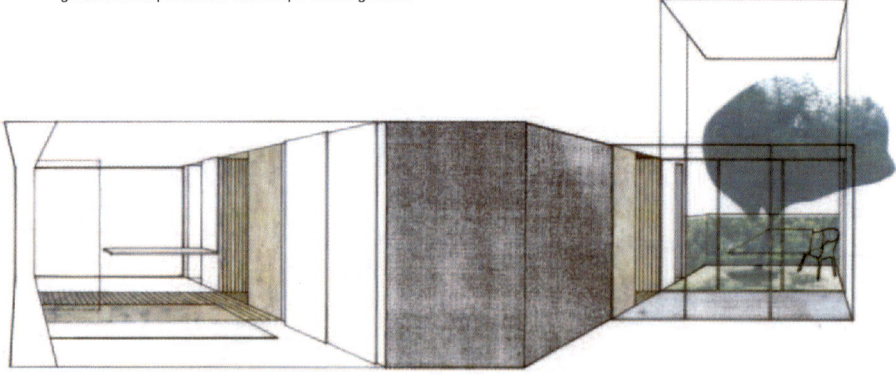

Plate 13 and Plate 14

Urban Nature Towers:
A proposal for a high density CPUL accommodating 450 persons per hectare. Vertical landscaping, created by attaching a framework to the building's façade, supports trained soft fruit plants and fruit trees. The vertical landscape supplements horizontal urban agriculture fields.

Plate 15

CPULs architecture London Shoreditch

Ariel view of the Urban Nature Tower site. A linear public park (light green in the image) surrounds urban agriculture fields and is continuous with a larger urban green grid.

Plate 16

25

MORE CITY WITH LESS SPACE: VISION FOR LIFESTYLE

Katrin Bohn and André Viljoen

CARROT AND CITY: PRACTICAL VISIONING

VARIETY OF OCCUPATION AND OCCUPANTS

An important characteristic of CPULs is the way in which a variety of occupations occur, such as gardening, farming, commuting, playing sport, leisure time activities like parties and picnics, which are undertaken by a variety of occupants, for example, schoolchildren, market gardeners, city dwellers, retired people. . . . This variety of occupants may engage with one or more of the occupations found within CPULs. The range of possible permutations between an individual occupant of a CPUL and their single, or many, activities or occupations is large and greater than in many public facilities, such as leisure centres. CPULs combine the tranquil qualities of a park with physical activities. They are as likely to be occupied by someone seeking a place to rest and read, as by someone else wanting physical exercise (Figure 25.1).

Continuous networks or local areas of productive urban landscape will provide the space for a variety of activities. Open-air sports areas can be imagined, not bounded by a fence, or enclosed by a roof, but rather a loose network of paths for running and fields for games, like an open-air gym. Schools, leisure centres and clubs overlooking Continuous Productive Urban Landscapes gain access to a shared public realm, providing external space for a number of activities and occupations.

Figure 25.1

Figure 25.1 *LeisurESCAPE CPUL, London: Urban edge buildings overlook a productive landscape.*

Proximity to 'nature', reducing commuting distances and persistent visual stimulation, these qualities, enriching to the experience of place, all suggest the development of building types which engage with a new overlap between green space, dwelling, and use. At a modest scale Figures 25.2 and 25.3 show a proposal mixing spatial types, occupation, building and landscape. Double height balconies create space for landscaped 'interiors', and rooftops are crowned with bristling vegetation (see Plate 14). Paths and tracks on roofs reinvent the ideal of the Marseilles Unite.

ECONOMIC RETURN FROM LAND-USE

Economic return from ground-use can be measured in two ways: one way is to measure direct economic benefits resulting from new employment and enterprises; and the other, arguably more important in the long term, is to measure reductions in environmental degradation, due to productive urban landscapes. These benefits, accruing from reduced environmental impact, lessen the future costs associated with remedial environmental work.

We use the term 'site yield' to refer to quantifiable environmental benefits resulting from sustainable development, which utilises, for example, renewable energy systems, rainwater harvesting or includes urban agriculture. Site yield records the proportion of energy, and food requirements that can be harvested from within the boundary of a particular site. In order to get an idea of site yield for developments built to high standards of energy efficiency, we have annualised our design proposals for actual sites in England (Sheffield, Newark and Shoreditch). The density of occupation for these three sites varies between 92 and 450 persons per hectare. Figure 25.4 indicates the arrangement of buildings and open space on particular sites: Newark consists exclusively of dwellings; Sheffield can accommodate living and working; and Shoreditch provides accommodation for dwellings and social functions, such as libraries, sports faculties and schools. Figures 25.5, 25.6 and 25.7 show that fruit and vegetable production, within the boundary of a site, can be expected to account for about a quarter of annual requirements for densities at or below circa 200 persons per hectare. With densities as high as 450 persons per hectares, if only horizontal surfaces are used for food growing, then fruit and vegetable yield drops to about 10 per cent. By introducing a system of vertical landscaping to high density schemes, fruit and vegetable yields may be increased to circa 30 per cent of annual requirements. In all cases, the proposals are for buildings designed to high standards of energy efficiency, and insulated to levels which effectively eliminate the need for space heating in the English climate (Viljoen, 1997 and Viljoen & Bohn, 2000). Accommodation maximises benefits from natural ventilation and daylight. Buildings use roof space for accommodating solar hot water panels, for the supply of domestic hot water for washing, and photovoltaic panels for the generation of electricity to offset annual domestic requirements. Results from these studies show the significant contribution solar systems can make to energy requirements. The results also suggest that density will limit the potential for developments utilising solar power for all energy requirements when it approaches 450 persons per hectare.

The potential for collecting rainwater from roofs, for domestic consumption, is modest for all three examples. Organic urban agriculture has an indirect

CARROT AND CITY: PRACTICAL VISIONING

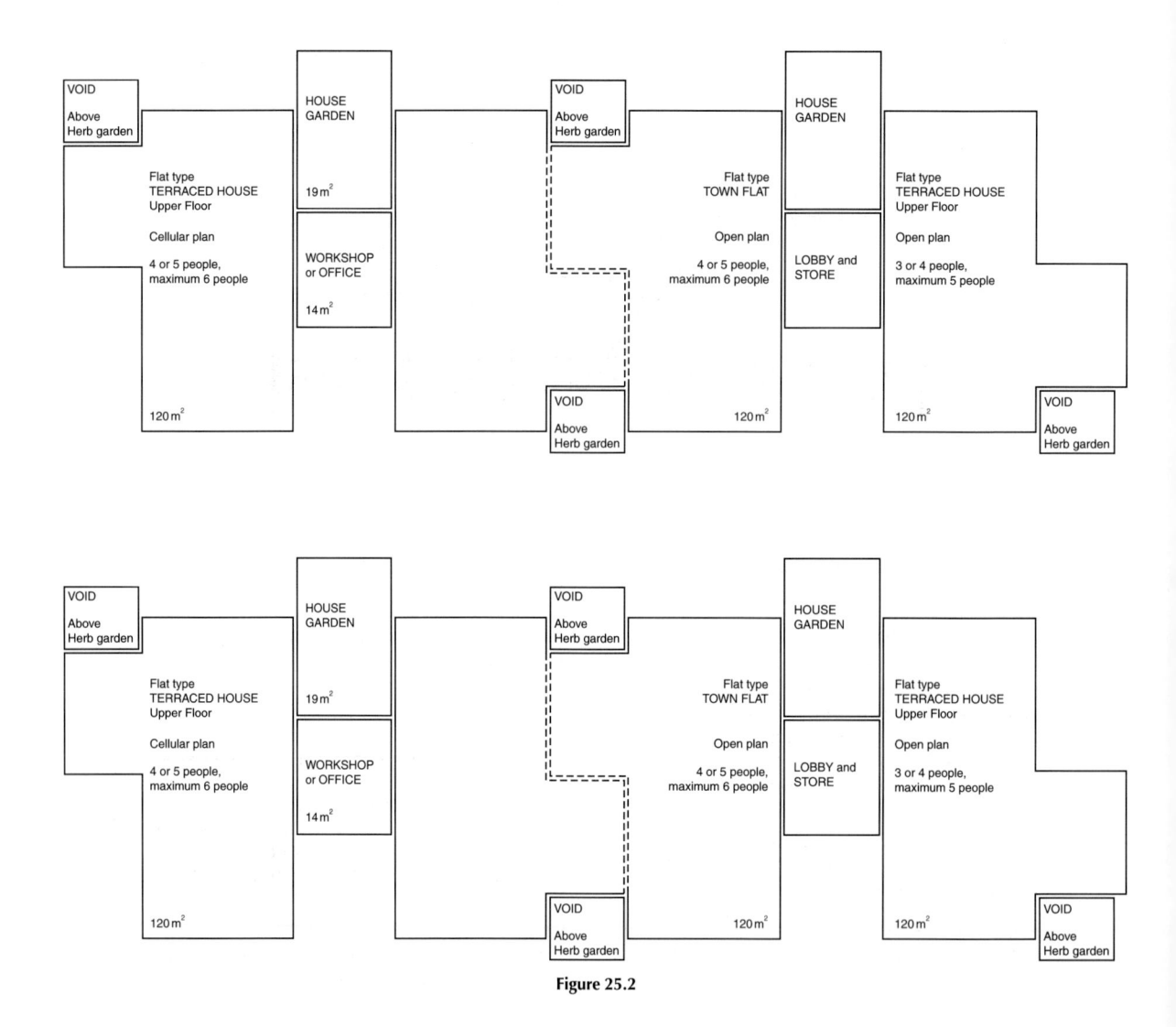

Figure 25.2

Figure 25.2 *Victoria Park CPUL: Apartment plans showing private 'house garden' balconies, workshop and office space adjacent to public and productive landscape.*

MORE CITY WITH LESS SPACE

Figure 25.3

benefit with respect to reducing water consumption. Enriching soil with composted organic matter improves its water retaining potential, thereby requiring less water for irrigation.

Figures 25.5, 25.6 and 25.7 represent ecological yields derived from particular sites, integrating a number of environmental features in a coherent overall design strategy. Significant additional contributions to the percentage of locally produced food would result from the development of peri-urban agriculture.

Instigating a bounding ring, or wide network of market gardens adjacent to densely built urban centres, provides further opportunity to reduce dependence upon remotely produced food. Considering that allotments were providing 50 per cent of Britain's fruit and vegetable requirements during the Second World War (Crouch and Ward, 1988), it is not impossible to imagine a large city like London aiming for self-sufficiency in fruit and vegetable production from CPULs and peri-urban agriculture.

Further indirect economic advantages will be derived from health and social benefits (see Chapter 3).

Direct Economic Benefits

One of the first questions asked about the viability of urban agriculture in existing cities is: where will the land come from? The answer will depend on whether the urban agriculture is to be located within the built-up area of an existing city or if it is to be sited within a planned urban extension on greenfield or brown-field sites (see Figure 25.8).

Greenfield and brownfield sites provide extensive areas of land within which CPULs may be developed. For urban agriculture, the availability of open land is not the only requirement – soil type and

Figure 25.3 *Victoria Park CPUL: External view of apartments and landscape.*

Sheffield site plan
92 persons per hectare

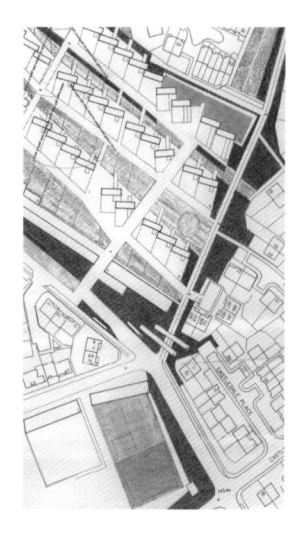

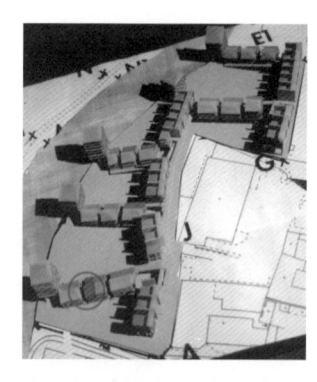

Newark site model
214 persons per hectare

Shoreditch site model
450 persons per hectare

Figure 25.4

condition will have a significant bearing on where crops can be grown and the degree of soil repair that may be required (see Chapter 8).

Integrating CPULs into existing cities will require the consideration of a number of different sources for land: existing undeveloped land; land from sites which are due for redevelopment; land within proposed developments; and portions of existing open space such as parks which may release pieces of land for CPULs. All recycled land, that is land which has had a previous use, will require testing for contamination and may require the application of appropriate techniques to 'repair' the soil.

Within existing cities roads provide an important source of land for CPULs. Converting roads into routes for connections between patches of open urban space and urban agriculture is technically relatively straightforward, converting them for use as urban agriculture sites is more challenging.

If roads are to be converted for crop growing, either raised beds or fields will have to be created. Both of these options will require the reintroduction or recreation of viable topsoil, which will have been removed during the construction of the road. The amount of road space given over to food growing will need to be considered in each situation, and issues such as the viability of breaking up and

Figure 25.4 *Comparative study of CPULs at different densities.*

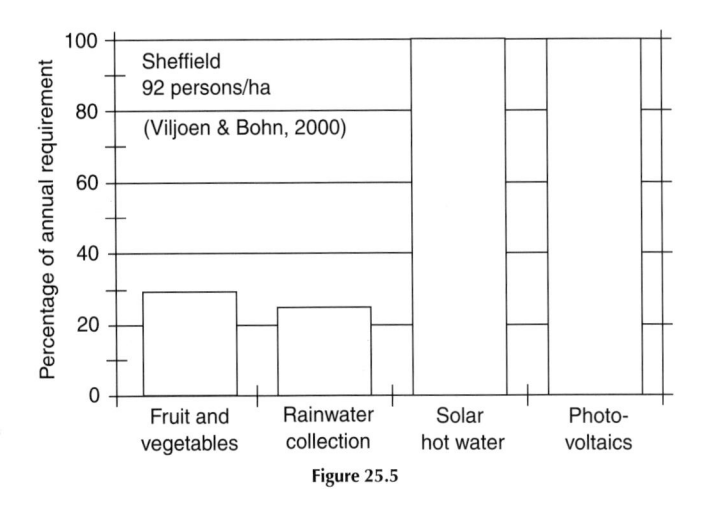

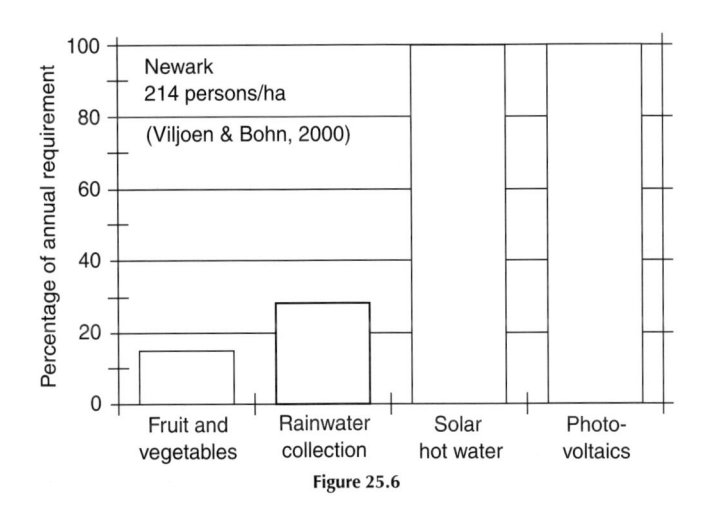

Figure 25.5 *Site yield for ElastiCity CPUL, Sheffield.*
Figure 25.6 *Site yield for Newark CPUL.*

Figure 25.7

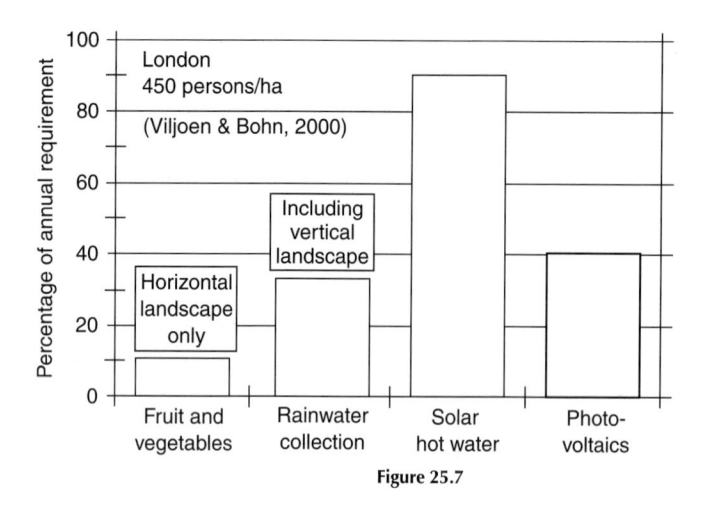

crushing materials judged against the many returns given by urban agriculture. In many European cities the complete demolition and onsite crushing of buildings, paving and parking areas is now common practice when a new use cannot be found for them. The crushed materials are then reused as aggregate or fill in new developments. In London, extensive industrial areas to the east of the city have undergone such a transformation since about 2000, and this process can equally be applied to selected roads.

If roads are converted to CPULs, a city or developer will not have to purchase new land, as it is already there. Furthermore, the introduction of urban agriculture, with its associated pastoral and ornamental qualities, may increase the value of adjacent land, commensurate with the quality of life improvements for residents inhabiting developments alongside the landscape. These economic benefits are like the indirect benefits derived from Amsterdam Plein, or Trafalgar Square, each having their own costs, but with collateral benefits for adjacent spaces.

Brownfield sites provide another source of land in existing cities. Where these have previously had an industrial use, the ground is likely to be contaminated and require detoxification (see Chapter 8). The techniques required to detoxify contaminated soil are now well understood and frequently applied. The costs are relatively high and for this reason it is often confined to sites that have the potential for high economic return once developed. Inspection of proposals for new residential development on land to the east of London, in the 'Thames Gateway' shows that a significant proportion of the site area of land for residential development is not built on. This land, within the curtilage of brownfield sites, could be used for productive uses in a manner similar to schemes described in this and the previous chapter (see Figure 25.4).

Figure 25.7 *Site yield for Shoreditch CPUL.*

MORE CITY WITH LESS SPACE

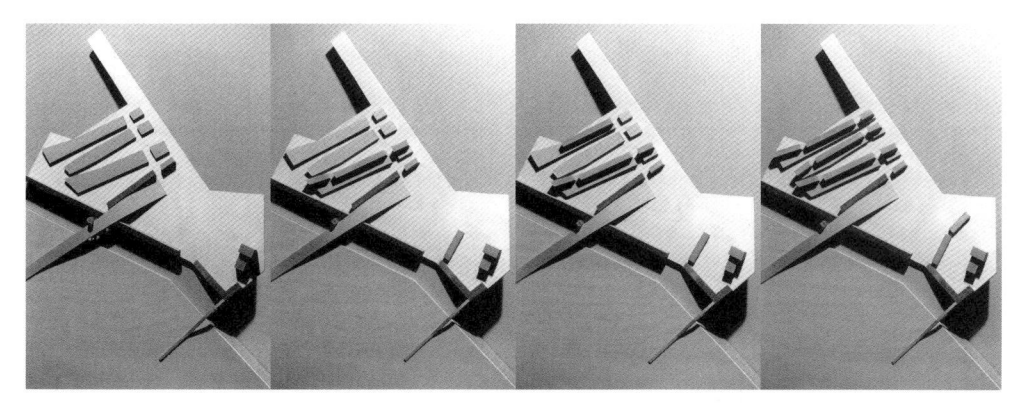

Figure 25.8

Figure 25.9

Figure 25.8 *ElastiCity CPUL, Sheffield: Site model showing incremental development. Urban agriculture fields structure the site and portions of these become plots for future buildings.*
Figure 25.9 *ElastiCity CPUL, Sheffield: External perspective showing the relationship between urban agriculture fields, garden/courtyards and dwellings.*

CPULs and urban agriculture in particular can be utilised as part of a strategy for incremental development. Incremental development links economic and environmental strategies. It accepts that development will not always occur at one time, indeed a site may take many years to reach its final form, or may even be conceived as a place that can continuously evolve and adapt. On the Manor Estate in Sheffield, we have proposed such a development. Here, an impoverished estate of terraced houses could be transformed into a CPUL, parts of which would erode over time as house building occurs, as and when funds and necessity permit. This strategy allows a cheap initial intervention on the site, which would mark plots for urban agriculture fields, paths and roads. Economics can determine when supporting infrastructures are installed. Figure 25.4 (see Sheffield site plan) illustrates the final form the development would take, with fingers of housing alternating with fingers of urban agriculture, connected to a larger continuous landscape. Initially, fields and roads are marked on the site, and the fields are operated as commercial market gardens, wildlife gardens, play and sporting facilities, providing an economic and environmental return from the land. As housing is built, some of the urban agriculture fields are given up and become plots for houses. Although this process accommodates unpredictable future scenarios, it provides a coherent framework in which development can occur. Building can progress quickly or slowly, but at any stage the place will read as a complete entity (see Figure 25.9).

Urban agriculture can also be understood as a means of supporting emerging enterprises, such as farmers' markets. Nina Planck describes how these have taken root in London (see Chapter 10). We are now in a situation where a network of markets, that is sales outlets, exist, but farmers are travelling on average a total of 100 miles (160 km) to and from markets (see Chapter 3). Currently, an infrastructure for the sale of locally grown food exists, but the supporting infrastructure of adjacent market gardens does not. The environmental case for local food production has been made and the economic desire for fresh locally grown food is evident in cities as disconnected from their productive hinterland, as is London.

Another economic return of CPULs will be that workplaces and dwellings can be sited within an environment of natural or rural characteristics, thereby answering those human desires, which are so well illustrated by the popularity of commuter belt dwellings located on the urban/rural fringe. CPULs will offer an inner-urban alternative to suburban sprawl, reducing drastically the commuting required between dormitory suburbs and urban centre (see Figure 25.10).

INNER-CITY MOVEMENT

In London, this is achieved by identifying existing patches of un- or under-developed land, of parks or playing fields and then planning and designing their inter-connection. Generally, roads provide the connecting element. Careful consideration of access requirements and circulation patterns usually indicates a number of roads that may be closed to through-traffic (see Plates 6, 7 and 8).

In London, for example, a cycle path running all along a particular CPUL would allow a person living in East Croydon (circa 20 km south of the city centre) to reach the city centre by bicycle in about an hour.

The fields proposed for the Manor Estate in Sheffield (see Figures 25.8, 25.9 and 25.10) provide a landscape resource for the city of Sheffield,

MORE CITY WITH LESS SPACE

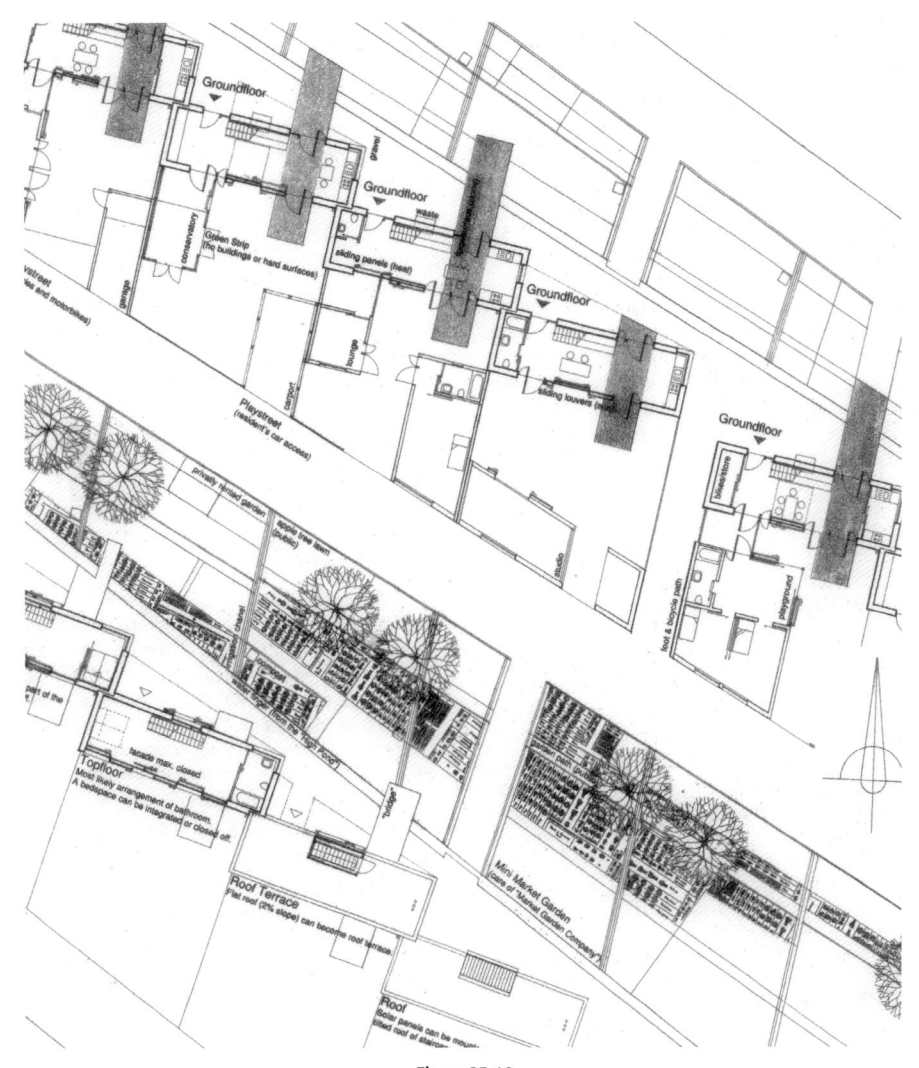

Figure 25.10

Figure 25.10 *ElastiCity CPUL, Sheffield: Plan showing the relationship between internal and external space. A play street mediates between houses and productive landscape, urban agriculture forms a changing 'sculpted ornament'. Visual and material connections are made between interior and exterior.*

allowing people to wander through and between urban agriculture sites and activity fields. As houses are built, the productive urban landscape provides a context into which they are placed. Contrast this with much current development on the edges of cities or on brownfield sites, where new houses are built adjacent to derelict wasteland. In this proposal for Sheffield a two-way partnership is established. New residents are in close proximity to a living landscape, like a large garden, or an emerging fragment of countryside. The city gains an evolving landscape, a place of intrigue for the curious, a place, which because of its evolving character, can become a focus for discussions about future urban development. By being an active place, it will provide a catalyst for a communal debate about future development strategies.

Plate 7 shows the large number of existing parks and under-used patches of open space which, with a relatively modest number of road transformations, could allow one to move unimpeded across London within a landscape with rural characteristics, a landscape of urban nature.

As a network of productive urban landscapes develops, and grows, so accessibility is enhanced in two directions. Paths are developed which can extend from the city centre outwards, towards fields of urban agriculture and in the opposite direction, paths develop, leading from suburban fringes into the city centre. Figure 24.1 from the previous chapter shows an area of land adjacent to Victoria Park, which despite its proximity, was physically and visually isolated from access to landscape. In the proposal, we see how landscaped fingers can create a physical connection between different-sized fields of urban agriculture, a park and dwellings. Thus, boundaries become blurred and in this way positive qualities of one condition are introduced to the other condition. So the suburb gains access to the theatre, and the central business district gains access to the landscape. We can claim with some certainty that these benefits will arise, as the shortcomings of each are shortcomings of omission, rather than an inherent weakness in the qualities that already exist in a particular location. For example, suburbs offer generous open spaces, views to the horizon and access to sunlight, whereas city centres thrive due to a compact arrangement of cultural and social venues. A productive urban landscape set in a network of continuous landscape provides direct connections, supplemented by public transport, between these different areas and activities. Suburban dwellers gain a pleasant walk to work. City dwellers get a walk into the countryside, like a weekend escape.

ENVIRONMENTAL DELIGHT

With the introduction of CPULs, habitat for animals and birds will increase and therefore so will biodiversity, an example of Ecological Intensification. At the same time, the development of composting systems in support of organic urban agriculture will improve soil condition. Sight and sounds within the city will change. Composting will reduce the number of refuse trucks and improved biodiversity will reintroduce the dawn chorus (see Figure 25.11) with the sound of birds, and insects.

In the earlier paragraph describing exposure to nature, we see how a number of natural phenomena and practical requirements overlay a site; each condition animating the place in a different way, each providing inhabitants with a different meaning for

MORE CITY WITH LESS SPACE

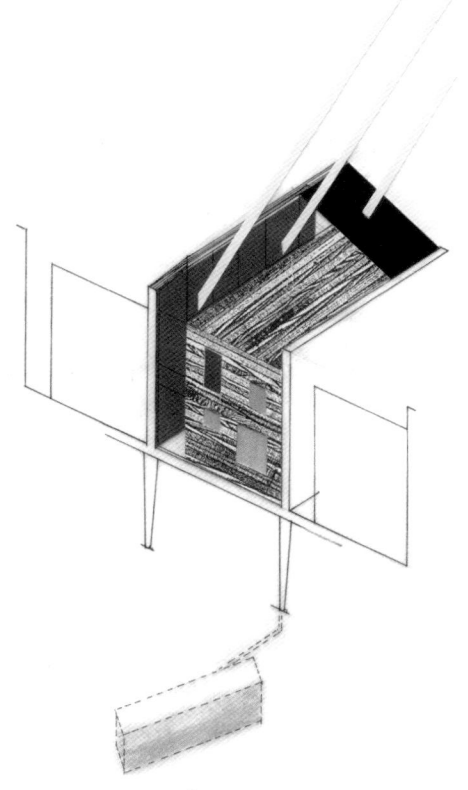

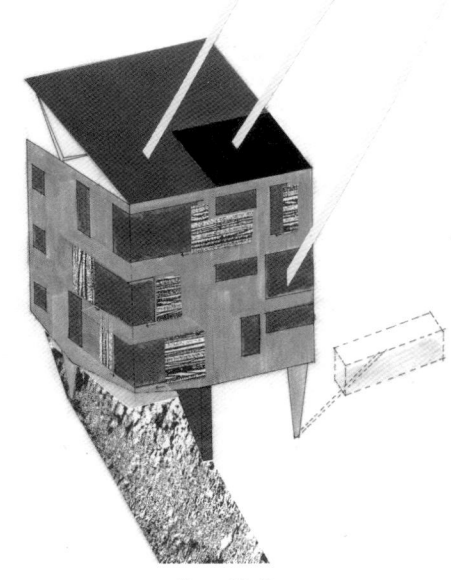

Figure 25.12

Figure 25.11

Figure 25.11 *Newark CPUL: An individual house designed to minimise non-renewable energy consumption. Its external envelope is super-insulated and natural energy systems are harnessed for domestic energy requirements. The roof collects solar energy and rainwater while also providing private external living space.*

Figure 25.12 *Newark CPUL: In this low energy apartment building, the roof collects solar energy for domestic hot water heating and for electricity generation. Rainwater is collected for domestic use. Building and landscape are responsive to the environment.*

CARROT AND CITY: PRACTICAL VISIONING

Peri-urban agriculture

High density compact urban core

Urban agriculture

Decompacted urban fabric

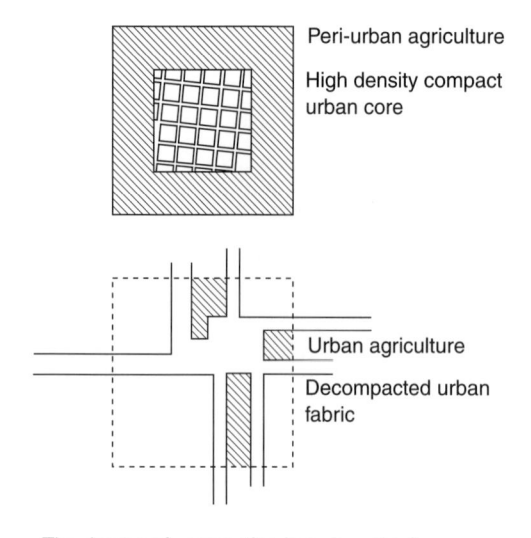

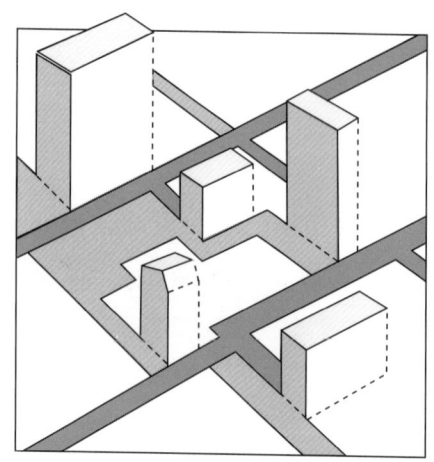

The degree of compaction in a city will influence the appropriateness of locating productive landscapes centrally or peripherally

Productive urban landscapes have the potential to create new networks of horizontal and vertical green spaces within cities

Figure 25.13

and experience of the place. This is like environmental choreography, marking and mapping experience in space and time. Water (rain) is collected, sunshine penetrates, occupies and conditions space, generates electricity and heats water (see Figures 25.11 and 25.12).

Agricultural fields create changing patterns or paintings in the landscape; this connection and celebration of the real generates delight in the environment.

REFERENCES:

Crouch, D. and Ward, C. (1988). *The Allotment*, Faber and Faber, London.

Viljoen, A. (1997). The environmental impact of Energy Efficient Dwellings taking into account embodied energy and energy in use. In *European Directory of Sustainable and Energy Efficient Building 1997*. James and James (Science Publishers).

Viljoen, A. and Bohn, K. (2000). Urban Intensification and the Integration of Productive Landscape. July 2000. *Proceedings of the World Renewable Energy Congress VI, Part 1*, pp. 483–488, Pergamon.

Figure 25.13 *Compaction, de-compaction and strategies for introducing CPULs.*

26

MORE OR LESS: FOOD FOR THOUGHT

André Viljoen and Katrin Bohn

The compact city model is currently favoured as that most likely to support sustainable development. Its major benefit in relation to environmental sustainability is the reduction in travelling distances and hence transport, due to compaction and mixed-use development. We see Continuous Productive Urban Landscapes (CPULs) complementing compaction, by including the major contribution urban food production can make towards environmental sustainabilty. Furthermore, we think that this combination will create a new kind of city, one with a richness of associations and experiences, to date found either in the city or the countryside. The CPUL model challenges the notion that all brownfield sites should be built upon, but does not challenge the principle that all land should be used to maximise its sustainable return.

The contemporary first world city is a facsimile for what and where it is not. Supermarkets, especially, manipulate people's perception of the availability and cost of food: everyday, all year round, there are aubergines and Parma ham, bamboo sprouts and pineapples, oysters, oranges, kebabs and samosas, chocolate powder and kiwi . . . each a reproduction of somewhere else. Such separation from geographical reality is fantastic – as a treat, but instead it has come to characterise and epitomise the city. This unsustainable model is based upon the outdated assumption that anything can be relocated from one place to another at any time.

The CPUL vision of the city is one which celebrates the material and the real, one which 'makes visible'. Within the contemporary European city many people are no longer conscious of the relationship between life and the natural processes required for its support. A CPUL city engages fully with elements such as a territory's seasonality, climate, weather, topography and vegetation. It is based on the ecological principles of life and the space required to accommodate all its actions, reactions and interactions. City dwellers have become passive observers of seasons (which they still often miss) or weather (which they often fear). The collective loss of environmental memory makes the natural context and the sequence of its processes less and less comprehensible. People are losing touch with the reality beyond their city boundary.

CPULs do not require the complete rebuilding or demolition of cities, rather they suggest reconfiguring the city so it can operate within the envelope of its own environmental capacity and as far as possible make its own ecological footprint equitable.

In this context, the vision of the city is one in which the resources required to support occupation become visible, imprinted on the urban tissue. A sustainable urban ecology becomes a key indicator for the successful city. It will be a city which, although complex, is comprehensible and flowing.

CPULs will be what other elements of urban infrastructure are now: they will be extensive and complex, demanding planning, management and maintenance. Like other elements of infrastructure, for example the electricity supply network, they will best be introduced incrementally.

At the same time, CPULs will be different to familiar urban infrastructures which mainly deal with distribution and circulation, like roads, railways, networks for energy supply, water and waste disposal. Though CPULs will also provide a network for circulation, there will be productive elements embedded within them, which add directly to our positive experience of the city. They will be environmentally and socio-culturally beneficial and economically viable. The range of experiences and lifestyles that a city can offer will be increased. CPULs will be networks that expand to accommodate occupation and production. This is unique.

CPULs will only be implemented if their vision is attractive and seen to be viable. These cases can be made, although the integration of CPULs will not be without difficulties and these should not be underestimated.

The English new town Milton Keynes provides one example of how CPULs might be funded, and the British Dig for Victory campaign during the Second World War shows that up to 50 per cent of fruit and vegetable requirements could be supplied by urban agriculture, although our own design research suggests that 25 per cent is a more realistic target for new development. Most of the remaining requirements could be provided by peri-urban agriculture. Urban agriculture will be a fundamental feature of CPULs, essential if the latter's environmental benefits are to be fully achieved.

The international examples of urban agriculture discussed in this book, although each dealing with a very particular set of conditions, identify a range of benefits from urban agriculture within CPULs, such as reductions in food miles, organic food production, creation of wildlife habitats, transport networks, educational resources, and an economic efficiency by concentrating intensive infrastructures in designated areas zoned for buildings.

Balancing benefits with a number of factors related to land and geography, such as size of the city, urban density, land ownership, soil type, climate or infrastructure, will determine where urban agriculture is and is not appropriate. This balancing act is a common feature affecting the development of any large-scale infrastructural project.

If Cuba can manage to implement a procedure as radical as its urban agriculture programme and sustain it for twelve years in conditions of economic stress, should it not be possible for any country to instigate programmes as comprehensive as those found in Cuba?

The Cuban programme for urban agriculture demonstrates how it can extend to the promotion of healthy eating, sustainable urban development or environmental education. Organic certification has not been introduced in Cuba, but the urban agriculture practised in Cuba is, by any practical measure, organic. Cuba demonstrates the viability of organic agriculture allied with intensive maintenance by farmers.

While Cuba provides a working model for the extensive integration of urban agriculture, it does not necessarily provide the ideal model for the distribution, location and connections between plots accommodating urban agriculture. The conditions of stress under which Cuba introduced urban agriculture mean that sites were chosen for entirely pragmatic reasons. The patchiness of urban agriculture found in Cuban cities can be taken as typical of the first stages of a programme to integrate urban agriculture. These isolated fields are comparable to exiting urban parks and gardens, which can be thought of as spaces with specific characteristics, with the potential of being bound into a CPUL. As yet there are no CPULs in Cuba. Introducing CPULs would provide a coherence and structure to otherwise isolated urban agriculture sites and create a framework for articulating the spatial and urban qualities inherent in urban agriculture, and fully utilising their benefits as routes for circulation, occupation and Ecological Intensification. Beyond the environmental benefits associated with urban food growing, urban agriculture can act as a catalyst for revealing and intensifying the occupation of under-utilised urban areas. In so doing, cities would gain benefits arising from the provision of adjacent open space, communication routes ideally suited to cycling and walking, moderation of the heat island effect and a landscape which allows people to comprehend their relationship with the natural environment.

A number of urban characteristics noted below can be attributed to urban agriculture, and these are all significant within CPULs.

The informal, self-regulated use of sites for personal food growing, such as huertos in Cuba or allotments in the UK, raise issues in relation to how these interface with planned or formal networks. The opportunities for engagement across edges or boundaries may differ from those suggested by larger commercially viable urban agriculture sites. Issues of privacy and seclusion may be more significant for small-scale private growing when distinct communities may become established. The relationship between *informal use* and *formal networks* is important for CPULs.

Urban agriculture fields often act as *bridging devices* between areas of different occupation. They do this by making a visible and physical bridge between two places. By doing so they often define disregarded or hidden places as space within the city.

The *occupation of edges* alongside urban agriculture fields is evident, for example, as a place to dry laundry, accommodate shops or provide climatic comfort zones. Taking these examples seriously, and imagining the transformations and opportunities inherent in them, provides inspiration for architectural interventions, which provide places for public or semi-public exchanges. The occupation of edges encourages a connection with a pastoral environment.

Urban agriculture gives *measure to a landscape*. The way in which ground for planting is often terraced, faceted and shaped to accommodate undulating ground, articulates and makes visible the underlying topography. The actual dimension of crops, and of beds, provides another gauge for measuring landscape and allowing an individual to locate and position themselves within a particular territory. This ability to read a landscape and locate oneself becomes critical as contemporary globalisation makes environments more uniform.

Urban agriculture sites exist as urban *climate and seasonal registers*, and due to their characteristic marking of the ground read as *urban ornament*.

The open spaces of a city incorporating CPULs will alter the physical landscape and the landscape of occupants and occupation. On the ground, cultivators will sculpt a new urban infrastructure, ever changing, but ever familiar, as crops come and go. Adjacent to this, a landscape of circulation and movement will appear, as the population traverses tracts of an agrarian landscape, and others play on ground adjacent to fields. Toilers and thinkers will be placed in a rediscovered adjacency, one which is not about destroying the city or conquering nature, but one that enriches both by acknowledging their interdependence.

More experience for less consumption!

MORE OR LESS

Figure 26.1

Figure 26.1   *Organoponico Pastorita, Cienfuegos 2004.*

CPUL CONTACTS COMPILED BY JAMES PETTS

City Farmer website – the first stop on the web for anyone interested in UPA. http://www.cityfarmer.org/

ETC – an NGO based in the Netherlands which co-ordinates RUAF (see below). http://www.etcint.org/

RUAF (Resource on Urban Agriculture and Forestry) – publishes the Urban Agriculture Magazine, runs various UPA programmes in the South, and operates an online source of information. http://www.ruaf.org/

FAO (Food and Agriculture Organisation) – the United Nations organisation with a wide range of UPA programmes. http://www.fao.org/

IDRC (International Development Research Centre) – runs its own UPA programme (Cities Feeding People) and acts as secretariat for the SGUA. http://www.idrc.ca/cfp

SGUA (Support Group on Urban Agriculture), a group of NGOs, international governmental organisations, and academics providing a forum for discussion and support of UPA including most of those listed here

Sustain – an alliance of over 100 different public interest organisations in the UK. Runs the City Harvest project. http://www.sustainweb.org/

TUAN (The Urban Agriculture Network). USA based Network, founded by Jac Smit, researching and promoting urban agriculture internationally.

US Food Security Coalition – a coalition of various groups and agencies in the US concerned with food security. http://www.foodsecurity.com/

INDEX

Accra (Ghana), 35
ActionAid, 20–1
added value, 26, 43
Adelaide (Australia), 38
affordable
  food, 49, 59, 218
  fruit and vegetables, 51
  public transport, 8
Africa, 22, 24, 36, 69, 83, 194, 196, 197, 198, 230
Agenda 20, *51*, 85, 126, 220, 222, 226
agrarian reforms, 136
agribusiness, 21, 28, 58
agricultural clinics and shops (*consultorios agrícolas/tiendas consultorio agropecuario*), 140
agricultural fields, 4, 12, 203, 240, 264
agro-chemicals, 41, 136, 215
  see also chemicals
agrotourism, 234
air freight, 42
air quality, 15, 53, 72, 122
allotment plot design, 209
allotment size, 208–9
allotments, 4, 12, 46, 53, 61, 66, 73, *75*, 81, 83, 85–6, 101, 125, 154, 218, 219, 255
  Britain, 207–15
  community, 83, 85–6, 129–31
  open green space, 125–7
  urban, 98–9, 104–5
  as urban landscape, 127–9
Amazon, 34, 36
American Community Gardening Association (ACGA), 106
Amsterdam Plein, 258
Angola, 195
animal feeds, 34, 36, 137
animals, 35, 41, 46, 70, 84, 87, 139, 262
appendicitis, 45
area one person can work, 153–4
artificial fertilisers, 34, 36, 74, 92, 191, 211, 226–7
Ashram Acres community garden, 57
Auckland (New Zealand), 235
Australia, 38, 189
autonomous and creative activity, 125

back gardens, 12, 208, 219
bagasse, 36
Barcelona (Spain), 81, 151
Bath (UK), 240
Becon Tree Organic Growers, 226

Beddington Zero Energy Development, 38
Beijing (China), 36, 97
Belgium, 21, *111*
benefits, 62, 63, 74, 127, 189, 207, 218, 230
  community, 83–5, 86, 87
  economic, 57–9, 66, 72, 73, 253, 255–60
  environmental, 21, 62, 106, 153, 240, 253, 267
  health, 59–60, 220
  horticultural, 92
  socio-cultural, 57
Berlin (Germany), 111, 113, 121
bicycles, 6, 11, 29
biodegradable waste, 90, 92, 231
biodiversity, 15, 21, 26, 44, *120, 122, 125, 207, 220, 262*
bio-fertilisers, 138
biofuel, 54
biological control, 138, 190
Birmingham (UK), 57, 99, 127
Borough (Southwark), 81
botanical garden, 60, 114, 118, 141
  Barcelona, *111*, 115, 119
  New York, 37
Botswana, 231
bottom-up approach
  see top-down/bottom-up approaches
Bournville model village, 99
box scheme, 38, 46, 58, 59, 70, 80
Bradley, Jennifer, 60
Brazil, 33, 34
bridges, 240, 245
bridging devices, 268
Bristol (UK), 38, 59
Bronx (New York), 37, 106
brownfield, 15, 16, 61, 62, *111, 246*, 255, 258, 266
buildings, 4, 6, 11, 22, 24, 36, 38, 81, 98, 101, 106, 115, 149, 151, 154, 177, 185, 186, 189, 219, 233, 240, 241, 242, 248, 253, 258, *263*, 267
Bulgaria, 233–4
Burkina Faso, 76

Cabeza, Héctor R. López, 151
Cairns Group, 46
Cairo (Egypt), 69
California, 37, 81
Canada, 22, 24, 33, 41, 61
Cape Town (S. Africa), 194
carbon dioxide (CO_2), 22, 33, 224
  emissions, 23–5, 42, 44, 122
Carrot City, 7, 8

272

car(s), 6, 9, 23, 30, 38, 42, 62, 100, 118, 215
certification, 80
character, 11, 220, 262
    landscape, 116, 126, 129, 130, 187, 218–19
    organic, 142, 143, 144
    urban, 139, 149, 185
cheap food, 207
chemicals, 22, 43, 44, 63, 80, 188, 189, 214, 226
    see also agro-chemicals
children's farms, 83
China, 34, 35, 53, 55
choice
    allocation of labour, 68
    crops, 210–12, 215
    food, 7, 41, 43, 46
    fruit and vegetables, 28
    lifestyle, 7
    space, 118, 122
    urban space, 53
Cienfuegos (Cuba), 149–51, 158, *180, 181, 183*, 190
circulation
    routes, 63, 173
    space, 100, 159, 202, 248, 260, 266–8
'Cities Feeding People' research and awareness programme, 193
city centres, 109
city farms, 12, 61, 80–1, 83–5, 86, 87–8, 208
city harvest project, *74*
city of short ways, 8
City of Tomorrow and its Planning, The, 99–100
climate change, 22, 24, 44, 45
comfort zone, 159, 268
commercial crops, 20
commercially viable, 70, 235, 268
communal orchards, 153, 188
communal ownership, 53
community development, 57, 218, 226
community food security coalition, 59
community gardens, 22, 57, 58, 59, 61, 83–5, 86, 87–8, 105–6, 114–15, 118–19, 161, 208,219
compact city, 266
compaction, 53, 210, *264, 266*
composting, 22, 90–1, 138, 142, 190, 191, 226, 262
composting toilets, 226
comprehensible, 21, 159, 246, 266
Consejo Popular Camilo Cienfuegos (Havana), 142, 188–90
Consultative Group on International Agriculture Research (CGIAR), 198
consumer ignorance, 41, 46
consumerism, 58

consumers, 21, 24, 28, 41, 42, 43, 44, 45, 46, 58, 59, 68, 70, 72, 79, 80, 81, 139, 147, 151
consumption
    energy, 22, 106
    food, 12, 23–4, 41, 46, 61, 142
    fuel, 30
    grains, 137
    home, 66, 68, 81
    local, 41, 46
    meat, 34
    resource, 15, 39
    seasonal, 25, 26–30
    self, 207, 235
    sugar, 60
    vegetables, 144
    water, 142, 253, 255
contaminated soil, 62, 63
contamination, 4, 21, 42, 62, 207, 256
continuous landscapes, 4, 11, *12, 13*, 16, 51, 58, 60, 105, *111*, 240, 245, 260, 262
conventional farming, 26
Cook, 22, 58
cooking, 24, 129, 220
co-operatives, 70, 136, 137, 140
countryside, 5, 21, 35, 53, 54, 98, 105, 122, 127, 137, 141, 143, 203, 240, 245, 246, 262, 266
Coventry Composter Newsletter, 91
covered houses (*casas de cultivo*), 140
Craul, Phil, 63
crime, 9, 53, 57, 58
crop rotations, 138, 191
crops, 20–1, 25–30, 36, 37–8, 41, 43, 44–7, 62, 63, 66, 70, 71, 80, 109, 126, 130, 136, 138–9, 141, 143, 144, 147, *148–9*, 151, 154, 158, *161, 163, 165, 167, 169, 171, 173, 175, 177, 179*, 188, 189, 190, 208, 210, 213–14, 219, 224, 226, 232, 245, 256, 268
crop species, 210
crop varieties, 22, 44, 210
Cuba, 29, 36, 68, 72, 83, 101, 136, 137, 138, 139, 140, 143, 144, 147–9, 151, 153, 154, 158, 188, 189, 191, 230, 232, 233, 240, 267–8
Cuban Academy of Sciences, 188
Culpeper community garden, 86
culture, 22, 53, 97, 105, 129, 130, 131, 136, 207, 215, 219
cycling, 6, 9, 127, 215, 224, 267

daily calorie intake, 23
Dar es Salaam (Tanzania), 35, 69, 197, 233
daylight, 248, 253
Dean City Farm (Merton), 80

273

INDEX

deforestation, 34, 193
Delft (Netherlands), 230–1
demand, 4, 28, 34, 46, 47, 53, 54, 61, 70, 71–2, 81, 97, 105, 125, 127, 208, 209
democratic landscape, 218
density, 8, 36, 53, 54, 62, 91, 128, 149, 247, 249, 253, 267
Department of Health (UK), 43, 49
derelict land and buildings, 81
design professionals, 218
Designer's Manual, 222
developed countries, 23, 25, 28, 33, 34, 35–7, 59, 60, 106
diabetes, 45
diet, 15, 45, 59–60, 215
Dig for Victory, 101, 104, 125, 267
discrimination, 57
distribution, 15, 20, 26, 29, 35, 43, 44, 46, 58, 71, 149, 151, 159, 190, 225, 233, 266, 267
domestic hot water, 24, 253
Doncaster (UK), 57
Durban (S. Africa), 194
Durham Civic Society, 130

Earth Summit (1992), 20, 23, 106
East Croydon (London), 260
eating patterns, 49
eco-houses, 37–9
ecological footprint, 33, *74*, 207, 266
ecological intensification, 6–8, 62, 101, 204, 207, 262, 267
ecology, 109, 110, 122, 159, 198, 224, 266
Economic and Social Research Council (ESRC), 61
economic benefit, 57
economic conditions, 198
economic development, 35, 220
economies of scale, 22, 41, 70
eco-systems, 39
edges, 98, 127, 129, 147, 149, 151, 158, *179, 184, 204*, 262, 268
edible ecosystem, 222
education, 58, 61, 84, 85, 90, 110, 136, 138, 143, 144, 226, 267
Ejército Rebelde dam, 141
elasticity, *243, 244, 246, 257, 259, 261*
El Paraiso Farmers Group, 189
El Pedregal Intensive Cultivation Garden, 189, 190
embodied energy, 12, 15, 22, 23, 24, 26, *28*
employment, 7, 15, 51, 58, 66, 69, 73, 84, 87, 104, 140, 147, 190, 253

energy, 8, 12, 15, 24, 25, 33, 34, 38, 42, 90, 105, 128, 193, 220, 222, 230, 247, 266
energy efficiency, 39, 106, 253
energy ratios, 26, 27
energy usage, 15, 23, 24, 25
England, 20, 23, 24, 26, 28, 38, 49, 50, 62, 75, 154, 201, 208, 213, 215, 218, 253
environmental awareness groups, 105
environmental benefits, 21, 62, 106, 240, 253, 267
environmental ethics, 105
environmentalism, 55
equitable development, 9
Essex (UK), 37
ethnic minority, 80, 84
ethnodiversity, 210
EU Landfill Directive, 90
Europe, 12, 15, 21, 24, 25, 28, 34, 38, 53, 55, 99, 100, 101, 106, 109, 128, 136
European cities, 98
European Commission (EC), 21
  see also European Union (EU)
European funds, 86
European Union (EU), 42, 79, 106, 224
exclusion, 51, 127
exercise, 60, 69, 125, 127, 207, 215, 231, *252*
exports, 21, 29, 42, 47
extension organisations, 138
external economic benefits, 72
externalities, 115, 122, 253, *255*

factors of production, 69
Fair Trade movement, 29
farm, 7, 12, 22, 29, 53, 54, 58, 59, 61, 66, 73, 75, 79, 80, 81, 83–4, 86, 87–8, 105, 136, 154, 188, 208, 214
farmers' markets, 225, 226, 260
farmland, 29, 33, 34, 36, 79, 99, 218, 230
farmworkers, 45, 46
Federation of City Farms and Community Gardens (FCFCG), 86, 87–8
fences, 158
fertiliser, 21, 23, 34, 36, 37, 38, 70, 80, 101, 122, 137, 138, 143, 188, 191, 211, 215, 224, 226, 227
floral *organopónicos*, 153, 158
  see also organopónicos
Florence (Italy), 35
flowers, 35, 37, 80, 118, 139, 153, 232
Foeken, 193, 195
food access, 49, 51
Food and Agriculture Organisation
  see United Nations Food and Agriculture Organisation (FAO)
food chain, 41–7

food distribution, 22, 29, 41, 43
food growing, 12, 35, 36, 57, 58, 59, 60, 61, 62, 63, 69, 83, 84, 85, 87, 97, 99, 100, 101, 104, 105–6, 188, 208, 217–20, 227, 233, 235, 253, 256, 267, 268
food imports, 21, 71, 137
food insecurity, 35, 68, 69
food miles, 22, 29, 35, 40–7, 79, 267
food producers, 41, 59
food production, 12, 15, 20, 22, 24, 25, *27*, 28, 33, 35, 37, 38, 39, 45, 46, 49, 51, 57, 58, 60, 61, 62, 63, 69, 73, 75, 85–6, 92, 97, 98–101, 104, 105, 136, 137, 139, 143, 144, 147, 151
food retailing, 21, 22, 58
food scarcity, 68
food security, 22, 35, 44, 45, 59, 68, 72, 73, 76, 138, 144, 193, 198
foreign exchange, 20, 45
form, 22, 110, 127, 128, 147, 260
formal/informal, 59, 60, 66, 69, 70, 71, 83, 84, 105, 109, 127, 131, 196, 197, 233, 268
France, 21, 33, 35
free range slug patrol, 224
fresh produce, 28, 42, 43, 59, 69, 143, 218
fruit and vegetables, 6, 7, 25, 26, 28, 29, 30, 37, 45, 49, *50*, 51, 59, 60, 66, 79, 101, 125, 153, 214, 219, 225, 231, 257, *258*, 267
production, 188, 253, 255
Fundacion Antonio Núñez Jiménez De La Natualeza Y El Hombre (Havana), 188, 189
fungible income, 69, 76

Gabrone (Botswana), 231–3
garden, 33, 36, 37, 38, 46, 51, 53, 54, 55, 57, 59, 60, 61, 63, 66, 69, 83, 84, 85
Garden Cities of Tomorrow, 99, 100
Garnett, 57, 58, 66, 97, 106, 207, 209, 210, 213
Geddes, Patrick, 55
genetic base, 44
genetic engineering, 21
genius loci, 4
Geographical Information Systems (GIS), 49–50
Georgia (US), 80
Ghana, 36
Glenn Valley (Gabarone, Botswana), 232
global warming, 15, 22, 24, 42
grass-roots activity, 218
Greater Havana Fresh Vegetable Company, 190
green belt, 54, 99, 100, 127, 141
greenfield sites, 15, 62, 246
greenhouse gases, 22, 24
greenhouses, 28, 36, 37, 38, 70, 210, 226
green infrastructure
    see continuous landscape

green lungs, 9, 11
green manure, 138
gregarious space, 202
groundscrapers, 101
Growing Food in Cities Report, 58
Growing in the Community, 125, 126, 127, 130
growing schools, 85
growing season, 28, 36
GRUPHEL programme, 27, 36

habitat, 21, 54, 118, 220, 230, 262
Harare (Zimbabwe), 194, 198
Hart, Robert, 224
Havana (Cuba), 35, 36, 37, 70, 81, 135–45, 147, 149, 151, 153, 154, 156, *163, 169, 180, 182, 183, 187*, 188, 189, 190
Havana's Green Belt, 141
health, 8, 15, 21, 22, 37, 43, 45, 49, 51, 59, 60, 72, 73, 76, 80, 84, 86, 88, 110, 141, 151, 207, 213, 218, 220, 226, 227, 255
healthcare facilities, 137, 142
Health of the Nation Report, 59
healthy food, 49, 50
heart disease, 45, 59
Heathrow Airport, 33
Heeley City Farm (Sheffield), 84–5
Heeley Development Trust, 84
Henry Doubleday Research Association (HDRA), 90
high density urban living, 128
high efficiency organic orchard (Organopónico de Alto Rendimiento, OAR), 140
    see also organopónicos
home gardens, 69
horizontal intensification, 147, 230, 233
hormone disruptors, 227
hospitals, 73, 143
household budget, 72
Howard, Ebenezer, 99
Hubli-Dharwad (India), 67, 69
human scale, 147
Hynes, 57

imported food, 42
imported soil, 151
income, 37, 49, 58, 66, 68, 69, 70, 71, 72, 76, 79, 81, 86, 107, 110, 116, 207, 209
indices of access to food, 49
Indonesia, 34
Industrial Revolution, 97, 98–101
inequalities in health, 51
infrastructure, 45, 53, 70, 71, 72, 73, 75, 139, 140, 141, 142, 144, 147, 188–9, 214, 231, 240, 260, 266, 267, 268

INDEX

integration, 8, 12, 26, 61, 100, 110, 136, 141, 143, 144, 153, 197, 198, 242, 267
intensive farming, 188
intensive orchards (*huertos intensivos*), 140, *163, 182*
interconnectedness, 222, 224, 226
International Development Research Centre, Ottawa, Canada (IDRC), 193, 198
International Potato Centre (CIP), 198
Irkutsk (Siberia), 36
irrigation, 38, 141, 142, 190, 211, 255
Italy, 33, 35, 38, *111*

Jackson, 58
Japan, 34
Jakarta (Indonesia), *67*, 70
jobs, 57, 66, 69, 74, 143, 197, 198
Johannesburg (S. Africa), 194
Johnson, Samuel, 20

Kathmandu Valley (Nepal), 230–1
Kenya, 20, 36, 41, 193, *196*, 198
kindergartens, 143
kitchen gardens, 98
Kitilla, 233
Kowloon (Hong Kong), 53
Kramer, K.J., 24
Kyoto protocol, 25

land contamination, 62, 213
landfill, 44, 72, 90, 92
land property pattern, 137
landscape bridge, 245, *246*
landscape strategy, 11, 16, 153, *173*
landscaping industry, 92
landscrapers, 101
land-use planners, 61
land-use policy, 60–1
larger suppliers, 22
layout, 11, 16, 109, 110, 112, 116, 127, 129, 147, 149, 151, 202, 219
Leach, G., 23, 26,
Le Corbusier, 99, 100, 101
Leigh Court, 38
leisure-gardens, 105
LeisurEscapes, 233
Lenin Park (Havana), 141
Lesotho, 195, 196
Letchworth (UK), 99
lifestyle, 7, 8, 11, 33, 105, 109, 141, 215, 224, 266
Lincolnshire (UK), 225
linear industrial city, 100

Living City, The, 100–1
living landscape, 130, 262
local authority, 59, 61, 83, 86, 197, 219, 220
local demand, 47, 97, 209
local distinctiveness, 57, 220
local economies, 20, 45, 58–9, 72, 207
local food, 7, 15, 29, 33, 41, 44, 45, 46, 51, 58, 73, *74*, 81, 147, 154, 222, 260
local food activists, 79
Local Food Links, 59
local growing and trading of crops, 25, 29–30
local market, *14*, 29, 57, 59, 75
London, 4–9, 15, 29, 33, 37, 38, 50, 53, 54, 66, *74*, 81, *103*, 105, 106, 207, 208, 211, 225, 226, 240, 242, 255, 258, 260, 262
London Farmers' Markets, 100
long-term benefit, 109
low energy buildings, 62
low incomes, 50, 60

McHarg, Ian, 55
macro-economic aspects, 72–3
Madhyapur Thimi Municipality (Nepal), 231
Malawi, 196, 198
Malaysia, 34
Manor Estate (Sheffield), 260
manual workers, 101, 125
market entry, 71–2
market gardens, 12, 51, 54, 63, 66, 153, 158, *187*, 234, 249, 255, 260
Massachusetts (US), 80
materiality, 147
Mazingira Institute (Kenya), 193
measure, 8, 20, 33, 67, 110, 114, 118, 122, 253, 267, 268
medical centres, 188
Mediterranean, 34
mega-cities, 35, 67
mental health, 60
Metropolitan Horticulture Enterprise Havana, 142
Metropolitan Park (Havana), 141
Mexican growers, 81
micro-economic aspects, 68
micro-forests, 141
middlemen, 45–6
Mile End Park (London), 106, *111, 112, 113, 116, 117*
Milton Keynes (UK), 106, 107, 267
mini market gardens, *5*, 235
Mishev, 234
Mlambo, 235
Mollison, Bill, 222, 225
monocultures, 41, 43

276

Moulsecoomb Forest Garden and Wildlife Project, 201
Mozambique, 195, *196*
multi-cropping, 188
municipal park, 98, 106
municipality, 139, 151, 188, 231, 234

Nairobi (Kenya), 198
Namibia, 195
national alternative agricultural model (NAAM), 137, 138
National Group for Urban Agriculture (GNAU), 139, 142
nature, 20, 54, 58, 61, *74*, 98, *118*, 122, 125, 159, 198, 203, 222, 223, 224, 245, 246, 249, 253, 262, 268
Naturewise (London), 226
Nazeing (UK), 37
negotiated communities, 129–30
Nepal, 230–1
Netherlands, 24, 28, 42, 63, 213, 230, 271
networks, 100, 159, 201, 252, 262
   community, 83–4
   distribution, 29
   food, 59
   street, 49
   transport, 225
Newark (UK), 243–7, 253, *256, 257, 263*
new towns, 99, 106
New York, 37, 106
night soil, 36
nitrous oxide, 24, 25, 226
noise, 9, 11, 15, 22, 45, 122
non-governmental organisations (NGOs), 66, 193, 198
Novo, González M., 138, 144
nursery houses (*casas de posturas*), 140, 154
nutrient flows, 34–5
nutrients, 34, 35, 38, 71, 92, 138, 223
nutrition, 35, 45, 59, 143, 144, 193, 195, 198, 214, 227

obesity, 59, 60
Obudho, 193, 195
occupants, 11, 12, 16 109, 110, 112, 114, 116, 120, 121, 161, 201, 202, 244, 246, 252, 268
occupation, 110, 114, 252–3
oil shortages, 23
old people's homes, 188
open urban space, 8, 11, 12, 15, 16, 81, 98, 99, 105, 106–7, 109–12, 114, 116, 118, 120, 122, 256
opportunity cost, 69, 70
organic agriculture, 147, 224, 227, 267

organic farming, 26, 140, 191
organic food, *27*, 46, 79, 131, 143, 215, 267
organic material production centres (*centros de producción de materia orgánica*), 140
organic urban agriculture, 21, 25–30, 62, 143, 188, 190–1
organic waste, 22, 35, 72, 74, 90, 190
organopónico(s) 139, 140, 153, *154, 156,* 158, 188, 232
ornament, 35, 109, 130, 219, 232, 240, 249, 258, 268
Ouagadougou (Burkina Faso), 76
ownership, 53, 126–8, 131, 147, 159, 189, 209, 218, 219, 220, 267
ozone depletion, 42

packaging, 23, 25, 26, 29, 42, 43–4, 45, 73
parcelas, 140, 189
parks, 11, 12, 15, 34, 53, 54, 81, 87, 91, 98, 106, 107, 110, 112, 141, 189, 220, *234*, 246, 256, 260
   car, 15
   urban, 83, 92, 99, 105, 109, 267
Parliamentary Select Committee Inquiry into "The Future for Allotments", 125
patches, 12, 218, 240, 243
paths, 117, 128, 141, 154, 158, 177, *179*, 201, *202, 205*, 230, 232, 243, 244, 252, 253, 260, 262
peat replacement, 92
Peckham (London), 29
pedestrian and cycle routes, 16, 60, 177, 202
pedestrians, 16, 51, 177
people's council, 189
período especial en tiempo de paz (special period), 137, 141
peri-urban agriculture, 35, 36, 66–7, 100, 136, 138, 139, 151, *177*, 193, 197, 198, 230, 234, 255, *264*, 267
   East and Southern Africa, 193–8
   Havana, 140–5
   see also urban and peri-urban agriculture (UPA)
permaculture, 37, *161*, 188, 219, 221–7
pest and disease controls, 137
pesticides, 21, 25, 41, 42, 43, 44, 45, 60, 73, 74, 137, 138, 143, 189, 191, 211, 224, 226, 227
Phillips, Tom, 249
photovoltaic panels, 253
planning, 16, 37, 51, 52–5, 61, 62, 63, 73, *74*, 75, 85, 87, 99, 100, 109, 110, 126, 136, 143, 149, 151, 196, 197, 198, 215, 216, 218, 219, 224, 230, 231, 233, 234, 246, 260
planning guidance, 81, 151
plot size, 147, 208, 209, 240

INDEX

plots, 36, 37, 63, 66, 99, 101, 105, 126, 128, 129, 130, 131, 149, 151, 154, *165*, 190, 201, 207–10, 214, 215, 218, 219, 232, 235, 260, 267
policy, 36, 46, 49, 51, 60–2, 73–6, 80, 85–8, 126, 136, 140, 143, 146, 144, 190, 193, 196, 197, 198, 231, 233, 234
Policy and Planning Guidance, 81
policymakers, 39
politics at a small scale, 130
pollution, 9, 11, 73
  air, 4, 9, 22, 215, 226
  noise, 9, 45, *120*, 122,
  sewage, 34
  water, 37, 207, 211, 226
pollution absorption, 33, 92
polycultural planting (*huertos populares parcelas*), 140, 226
popular organic orchards (*organopónicos*), 37, 140, 143
Port Elizabeth (S. Africa), 194
poultry, 37, 66, 142
poverty, 20, 35, 49, 51, 58, 125, 193, 194, 197, 198, 209
poverty alleviation, 35, 193, 196, 197
PPG17 (Planning Policy Guidance Note 17: Sport, Open Space and Recreation), 126
Pretoria (S. Africa), 194
price, 4, 20, 21, 23, 26, 28, 44, 46, 62, 66, 69, 70, 71, 72, 73, *75*, 76, 79, 80, 107, 143
private car, 23, 62
processing, 22, 23, 25, 26, 29, 41, 42–3, 44, 71, 72, 73, 225, 231
production and producer partnerships, 136
productive microspaces, 130
productive potential, 139
productive urban landscapes, 4, 12, 49, 51, 63, 147, *154*, 156, 203, 222–7, 240, 243, 249, 252, 262, 264, 266
profitable, 26, 69, 214, 215
Programa Especial De Desarrollo de la Agricultura Urbana, 188
promotion, 26, 49, 60, 86, 92, 139, 196, 197, 234, 267
protect crops, 109
public health, 37, 44, 45, 60
public realm, 107, 126, 158, 219, 244, 252
public transport, 8, 49, 262

radio-concentric change-over city, 100
railways, 50, *74*, 98, 209, 266
rainwater, 6, 72, 211, 226, 253, *257, 258*
raised beds, 62, 63, 151, 154, 158, 189, 219, 256

recreation, 11, 38, 69, 83, 97, 106, 125, 137, 141, 142, *155*, 201, 209, 210, 230, 244, 245, 256
Rees, William, 33
renewable energy, 22, 23, 28, 105
Resource Centre for Urban Agriculture and Forestry (RUAF), 57, 235
resources, 15, 22, 33, 38, 39, 41, 49, 53, 58, 60, 68, 71, 73, 86–7, 91, 143, 151, 189, 197, 198, 220, 223, 224, 226, 231, 266, 267
retail deserts, 22
Rio de Janeiro (Brazil), 34
road freight, 41, 44, 47
roads, 6, 15, *49, 50*, 63, *116*, 122, 141, 149, *231*, 232, 240, 245, 256, 258, 260, 266
Rodas (Cuba), 149, 151, *152*, 153, 154, *167, 175, 181, 182*, 188
Rodgers, 22, 58
Rome, 34
roofs, 15, 211, 253
Royal Agricultural Society for England, 26
Rugby Borough Council, 91
rural, 21, 34, 35, 85, 91, 122
  agriculture, 37, 54, 71
  Cuba, 138–40
  land, 12
  way of life, 127
rural poor, 99
Russia, 36, 61, 66, 69, 72, 218
Rwanda, 195, *196*

St. Petersburg, 36
Sandwell (UK), 49–51, 219
schools, 6, 53, 58, 73, 84, 85, 87, 143, *156*, 188, 252, 253
Schrebergärten, 99
scientific networks, 139
seasonal consumption, 12, 25, 26–30
seasonal food, 29, 46, 60, 219
seasonal registers, 268
seasonality, 28, 46, 72, 149, 220, 245, 246, 266
second skins, 240
security, 22, 35, 44, 45, 59, 68, 70–4, 76, 126, 130, 131, 138, 144, 155, 193, 198, 226
selection criteria, 151
self-build, 126
self regulated, 201, 268
self-sufficient, 12, 20, 99, 188, 189, 231, 232
sewage, 34, 38, 231, 232
Shanghai (China), 35, 36, 70, 97
Sheffield (UK), 59, 84, *120*, 243–4, 245, *246*, 253, *256, 257, 259*, 260, *261*
Sheffield Environmental Training, 84
shelf life, 44

278

shoe leather costs, 71
shopper miles, 42
Shoreditch (UK), *247*, 248, 253, 256, 258
Shropshire (UK), 222
Silent Spring, 21
single regeneration funds, 86
sites of negotiation, 125
site yield, 253, *257, 258*
slow food movement, 22
Small Holdings and Allotments Act (1908), 99
social benefits, 255
social enterprise, 86
social inclusion, 83, 127
socialist block, 136
social participation, 83, 84, 85, 86, 87
Society of Friends, 101
Soil Association (UK), 59, 224, 225
solar gain, 248
solar hot water, 253, *257, 258*
Solidarity Park (Havana), 141
Sorvig, Kim, 63, 106
South Africa, 36, 194, *196*
space heating, 24, 253
spatial characteristics, 147–9
sports, 6, 7, 53, 83, 85, 86, 136, 253
sprawl, 9, *165*, 234, 260
Stanley, 25, 26
state owned self-production areas (*autoconsumos estatales*), 140, 154
structural adjustment policies, 68
structural barriers, 73
structure of soils, 92
subscrapers, 101
subsidised products, 20
suburbia, 8, 99, 234
suburbs, 53, 54, 79, 88, 235, 260, 262
supermarket, 21, 22, 29, 30, 37, 41, 42, 43, 44, 45, 46, 49, 57, 58, 71, 75, 83, 125, 214, 226, 235, 266
supermarket gardens
  see mini market gardens
surveys, 68, 158, 213
sustain, 59, 74
Sustainable Landscape Construction, 63
sustainable landscape design, 106

Tanzania, 36, 193, 194, 197, 198, 233
temporal rhythms, 249
tenure, 70, 73, 126, 198
territorial ordering scheme, 142
Texas (US), 81
Thailand, 34
Thames Barrier Park (London), 106, *111, 116, 117, 120, 121*

This Common Inheritance, 80
Thompson, J. William, 63
Thorpe, Harry, 104, 105, 128, 209
top-down/bottom-up approaches, 136, 144
topsoil, 256
Trafalgar Square (London), 258
traffic, 8, 11, 16, 34, 114, 118, 120, 122, 260
training, 57, 58, 61, 83, 85, 86, 87, 105 140, 194, *195*, 198
transnational corporations (TNCs), 46
tree-curtains, 141
trees, 12, 53, 92, 100, 107, 118, 128, 130, 141, 154, 188, 190, 222, 223, 240
Trojan (Bulgaria), 233–5

Uganda, *196*, 198
under-utilised, 71, 235, 267
unemployment, 68, 99, 101
United Kingdom, 21, 23, 24, 25, 26, 34, 37, 38, 41, 42–3, 44, 45, 46, 57, 59, 60, 61, 68, 70, 71, 73, 74, 80, 87, 90, 98, 101, 107, 128, 214, 218, 219, 220, 268
United Nations Environment Programme (UNEP), 45
United Nations Food and Agriculture Organisation (FAO), 21, 67, 196, 198
United Nations Framework Convention on Climate Change, 24
United States blockade, 136
United States of America (USA), 22, 24, 25, 26, 33, 37, 53, 55, 57, 58, 59, 61, 80, 81, 105, 106
University of Cienfuegos, 171, 177, 191
unplanned, 140, 219
Upper Bieslandse Polder, 230
urban agricultural reserve zones, 231
urban agriculture, 4, 11–15, 21, 22–30, 33, 35–9, 51, 53, 54, 55, 57, 58, 60–3, 66, 67, 72, *74, 75*, 91, 97, 100, 101, 106, 112, 122, 125, 127, 136, 138, 139, 142, 144, 201, 203, 207, 219, 220, 225, 226, 230, 231, 232, 233–5
  Africa, 193–8
  Cuba, 147–9
  Havana, 140–5
urban agriculture advice shops, 189
Urban Agriculture Magazine, 198, 235
urban and peri-urban agriculture (UPA), 66–7, 136, 193, 196, 207
  motives for, 68–9
urban and peri-urban agriculture (UPA) subprograms, 143
urban beaches
  see urban river fronts

INDEX

urban connections, 16
urban ecology, 188, 198, 266
urban economies, 72, 193, 196
urban farms, 22, 58, 59, 83, 105, 142
urban fields, 109, 246
urban food, 12, 21–2, 35, 36, 37, 38, 57–63
urban forests, 11, 109
urban fringe, 59, 69, 75, 149
urban landscape, 8, *11, 111*, 127–31, 218, 219–20
   productive, *12*, 14, 105, *113, 131*, 147, 213, 240, 245, 252, 260, 262
   regional, 8
urban metabolism, 38
urban parks, 83, 92, 99, 109, 267
urban regeneration, 57, 84, 85, 105
urban river fronts, 109
urban squares, 109
urban stages, 109
Urban Task Force, 128
Urban Vegetable Promotion Project (Dar es Salaam)197
usufruct, 70, 140
utility, 68
utility theory, 68

Vale, Brenda, 23
Vale, Robert, 23
Vancouver (Canada), 218
Venice (Italy), 16
vermi-compost, 191
vernacular tradition, 126
vertical fields, 248
vertical landscape, 240–1, 248
vets, 188
views, 4, 126, 171, *184, 185*, 243, 246, 262
Virginia (US), 79, 80
volunteers, 84, 86, 87, 201

Wackernagel, Mathis, 33
walking, 6, 9, 11, 54, 60, 99, 127, 169, 215, 249, 267
Washington, DC (US), 79, 81
waste, 21, 22, 38, 44, 53, 72, 231, 266
water, 4, 6, 15, 24, 36, 37, 38, 41, 72, 73, 74, 76, 92, *110, 114*, 120, 122, 140, 141, 142, 151, 190, 207, 211, 214, 215, 222, 226, 230, *232*, 253, 255, *257, 258*, 263, 266
Welbeck Road Allotments Association Trust, 85
Wellgate Community Farm, 87
Welwyn (UK), 99
Wessex Water, 38
western diet, 45
Whitefield, 222, 223, 224
wilderness, 11, 55, 118
wildlife habitat, 54, 230
Wimbledon Park, 80
window boxes, 218
Wisconsin (US), 80
World Bank, 20, 46
World Trade Organisation (WTO), 46
World War, First, 59, 101
World War, Second, 53, 101, *104*, 151, 207
Wright, Frank Lloyd, 100
Wuppertal Institute, 42

yield, *5*, 26, 27, 69, 71, 72, 112, 139, 147, *148–9*, 153, 154, 188, 190, 210, 222, 253, 255, *257*
Yoveva, A., 234

Zambia, 193, *196*
Zimbabwe, 193, 194, *196*
Zoo, 112, 113, 114, 115, 120, 121, 141